Underutilized Resources as Animal Feedstuffs

Subcommittee on Underutilized Resources as Animal Feedstuffs

Committee on Animal Nutrition

Board on Agriculture

National Research Council

NATIONAL ACADEMY PRESS
Washington, D.C. 1983

National Academy Press, 2101 Constitution Avenue, NW, Washington, DC 20418

NOTICE: The project that is the subject of this report was approved by the Governing Board of the National Research Council, whose members are drawn from the councils of the National Academy of Sciences, the National Academy of Engineering, and the Institute of Medicine. The members of the committee responsible for the report were chosen for their special competences and with regard for appropriate balance.

This report has been reviewed by a group other than the authors according to procedures approved by a Report Review Committee consisting of members of the National Academy of Sciences, the National Academy of Engineering, and the Institute of Medicine.

The National Research Council was established by the National Academy of Sciences in 1916 to associate the broad community of science and technology with the Academy's purposes of furthering knowledge and of advising the federal government. The Council operates in accordance with general policies determined by the Academy under the authority of its congressional charter of 1863, which establishes the Academy as a private, nonprofit, self-governing membership corporation. The Council has become the principal operating agency of both the National Academy of Sciences and the National Academy of Engineering in the conduct of their services to the government, the public, and the scientific and engineering communities. It is administered jointly by both Academies and the Institute of Medicine. The National Academy of Engineering and the Institute of Medicine were established in 1964 and 1970, respectively, under the charter of the National Academy of Sciences.

This study was supported by the Agricultural Research Service of the U.S. Department of Agriculture; the Bureau of Veterinary Medicine, Food and Drug Administration of the U.S. Department of Health and Human Services; and by Agriculture Canada.

Library of Congress Cataloging in Publication Data

National Research Council (U.S.). Subcommittee on Underutilized Resources as Animal Feedstuffs. Underutilized resources as animal feedstuffs.

Bibliography: p.
1. Organic wastes as feed. I. Title.
SF99.W34N37   1983   636.08′55   83-13311
ISBN 0-309-03382-9

Printed in the United States of America

Preface

In 1978, the Committee on Animal Nutrition (CAN) of the National Research Council (NRC) convened a task force of scientists to examine the available literature on the feeding of "exotic" or underutilized feedstuffs to food-producing animals and to recommend to the CAN whether a report on the subject would be warranted. Subsequently, a CAN Subcommittee on Feeding Underutilized Feedstuffs to Animals was appointed to include expertise in animal nutrition, animal waste recycling, food science and technology, crop residues, feedstuffs from wood, and biochemical engineering.

The subcommittee met on September 6–7, 1978, in Washington, D.C.; December 3–5, 1979, in Chicago, Illinois; and June 17–19, 1980, at the NAS Summer Studies Center, Woods Hole, Massachusetts.

This report was written by the CAN Subcommittee on Feeding Underutilized Feedstuffs to Animals, each chapter being prepared by one or two members of the subcommittee; however, the entire document has been approved by all members.

During the preparation of its report, the subcommittee received valuable assistance from Robert C. Albin; W. Brady Anthony; Dudley D. Culley, Jr.; Charles C. Dunlap; Marcel Faber; D. M. Graham; James F. Hentges, Jr.; L. D. Kamstra; John H. Litchfield; C. Reed Richardson; Lewis W. Smith; Jack C. Taylor; Peter J. Van Soest; Howard G. Walker; and R. J. Young.

Review of this report was accomplished through the advice and suggestions of the Committee on Animal Nutrition and the Board on Agriculture.

The subcommittee is indebted to Philip Ross, Executive Director, and Selma P. Baron, Staff Officer, of the Board on Agriculture for their assistance in the preparation of the report. The subcommittee is especially grateful to George K. Davis who served as coordinator for the review of the report.

Subcommittee on Underutilized Resources as Animal Feedstuffs

JOSEPH P. FONTENOT, *Chairman*, Virginia Polytechnic Institute and State University
ANDREW J. BAKER, USDA Forest Products Laboratory
ROBERT BLAIR, University of Saskatchewan
CHARLES L. COONEY, Massachusetts Institute of Technology
TERRY KLOPFENSTEIN, University of Nebraska
ROBERT C. PEARL, University of California, Davis
LARRY D. SATTER, USDA Dairy Forage Research Center

COMMITTEE ON ANIMAL NUTRITION

DUANE E. ULLREY, *Chairman,* Michigan State University
JIMMY H. CLARK, University of Illinois
RICHARD D. GOODRICH, University of Minnesota
NEAL A. JORGENSEN, University of Wisconsin, Madison
BERYL E. MARCH, University of British Columbia
GEORGE E. MITCHELL JR., University of Kentucky
JAMES G. MORRIS, University of California, Davis
WILSON G. POND, USDA Meat Animal Research Center
ROBERT R. SMITH, Tunison Laboratory of Fish Nutrition, USDI

SELMA P. BARON, *Staff Officer*

BOARD ON AGRICULTURE

WILLIAM L. BROWN, *Chairman,* Pioneer Hi-Bred International, Inc.
LAWRENCE BOGORAD, Harvard University
NEVILLE P. CLARKE, Texas A&M University
ERIC L. ELLWOOD, North Carolina State University
ROBERT G. GAST, Michigan State University
EDWARD H. GLASS, Cornell University
RALPH W. F. HARDY, E.I. du Pont de Nemours & Co., Inc.
LAURENCE R. JAHN, Wildlife Management Institute
ROGER L. MITCHELL, University of Missouri
JOHN A. PINO, Inter-American Development Bank
VERNON W. RUTTAN, University of Minnesota
CHAMP B. TANNER, University of Wisconsin
VIRGINIA WALBOT, Stanford University

PHILIP ROSS, *Executive Director*

Contents

OVERVIEW 1

1 INDUSTRIAL FOOD PROCESSING WASTES 5
  Introduction, 5
  Quantity, 5
  Physical Properties, 7
  Fruit and Nut Processing Wastes, 7
    Apple Processing Wastes, 7
    Citrus Processing Wastes, 9
    Peach Processing Wastes, 10
    Pear Processing Wastes, 10
    Fruit Canneries' Activated Sludge, 12
    Winery Wastes, 13
    Cacao Processing Wastes, 15
    Fruit Pits, Fruit Pit Kernels, Nut Hulls, and Nut Shells, 15
  Vegetable Processing Wastes, 16
    Potato Processing Wastes, 16
    Sweet Potato Processing Wastes, 21
    Tomato Processing Wastes, 23
    Other Vegetables, 25
  Animal By-Products, 28
    Dairy Whey, 28
    Seafood Processing Wastes, 30
    Poultry Processing Wastes, 31

viii   Contents

    Red Meat Processing Wastes, 32
    Alternative Uses for Food Processing Wastes, 38
    Animal and Human Health Problems and Regulatory Aspects, 38
      Pesticide Residues, 38
      Heavy Metals, 39
      Animal Health, 39
    Summary, 40
    Literature Cited, 41

2 NONFOOD INDUSTRIAL WASTES 46
    Introduction, 46
    Organic Chemical Industry, 47
      Quantity, 47
      Physical Characteristics, 48
      Nutritive Value, 52
      Processing, 54
      Utilization Systems, 58
      Animal and Human Health, 59
    Fermentation Industry, 59
      Quantity, 59
      Physical Characteristics, 60
      Nutritive Value, 60
      Processing Technology, 61
      Utilization Systems, 62
      Health Considerations, 62
      Regulatory Aspects, 62
    Municipal Solid Waste, 63
      Quantity, 63
      Physical Characteristics, 63
      Nutritive Value, 63
      Processing, 64
      Animal and Human Health, 64
    Research Needs, 65
    Summary, 65
    Literature Cited, 65

3 FOREST RESIDUES 69
    Introduction, 69
    Whole-Tree Residue and Fractions of Whole Trees, 69
      Quantity, 69
      Collectibility, 71
      Physical Characteristics, 73
      Nutritive Value, 74
      Processing Methods, 84

Utilization Systems, 100
Pulpmill and Papermill Residues, 101
  Quantity, 101
  Physical Characteristics, 106
  Nutritive Value, 107
  Processing, 110
  Utilization Systems, 110
  Sludges, 111
  Spent Sulfite Liquor, 113
Wood Residues as Roughage Substitutes in Ruminant Diets, 113
Animal Health, 114
Regulatory Aspects, 115
Research Needs, 115
Summary, 116
Literature Cited, 117

4 ANIMAL WASTES ... 121
Introduction, 121
Quantity, 122
Physical Characteristics, 122
Nutritive Value, 123
  Chemical Composition of Animal Wastes, 123
  Nutrient Utilization, 124
  Performance of Animals Fed Animal Wastes, 132
Processing, 144
  Ensiling, 145
  Dehydration, 145
  Other Processes, 147
Utilization Systems, 148
  Experimental, 148
  Industrial, 149
  Potential Utilization, 150
Animal and Human Health, 150
  Pathogenic Organisms, 151
  Harmful Substances, 156
  Quality of Products from Animals Fed Waste, 162
Regulatory Aspects: Federal and State, 163
Research Needs, 164
Summary, 165
Literature Cited, 166

5 CROP RESIDUES ... 178
Introduction, 178

x *Contents*

    Quantity, 178
    Physical Characteristics, 179
    Nutritive Value, 180
        Corn, 180
        Wheat and Other Small Grain and Grass Straws, 182
        Soybean, 183
        Grain Sorghum, 183
        Other Residues, 184
    Processing Methods, 187
        Corn, 187
        Wheat and Other Small Grains, 194
        Soybean, 196
        Grain Sorghum, 196
        Other Residues, 198
    Utilization Systems, 199
    Potential Utilization, 202
    Animal and Human Health Problems and Regulatory Aspects, 203
    Research Needs, 203
    Summary, 204
    Literature Cited, 204
    Glossary, 210

6 AQUATIC PLANTS 211
    Introduction, 211
    Quantity, 212
    Physical Characteristics, 212
    Nutritive Value, 214
        Chemical Composition, 214
        Nutrient Utilization, 214
        Animal Performance, 216
    Processing, 217
    Algae, 217
    Seaweed, 218
    Other Aquatic Plants, 219
    Utilization Sytems, 220
    Animal and Human Health, 220
    Regulatory Aspects, 222
    Research Needs, 222
    Summary, 223
    Literature Cited, 224

APPENDIX TABLES 228

Tables and Figures

Tables

INDUSTRIAL FOOD PROCESSING WASTES
1. Yearly Food Processing Industry Solid Residuals, by Product and Disposal Method, 6
2. Food Processing Industry Solid Residuals per Year, by Region, 7

NONFOOD INDUSTRIAL WASTES
3. Major Chemicals Produced in Liquid-Phase Reaction Systems, 49
4. Processes Identified as Possibly Having Underutilized NFI Waste, 53
5. Cell Conversion Yields on Various Substrates, 56
6. Protein Content of Various Microorganisms, 58

FOREST RESIDUES
7. Estimates of U.S. Aboveground Forest Biomass Potential, 70
8. In Vitro Dry-Matter Digestibility of Various Woods and Barks, 75
9. Water Solubility of Various Hardwood Barks and Extent of Carbohydrate Dissolution, 76
10. Feedlot Performance of Cattle Fed Pelleted Diets Containing Whole Aspen Tree Material, 78
11. Feedlot Performance of Cattle Fed Complete Mixed Rations Containing Alfalfa and/or Whole Aspen Tree Material, 80

xii   *Tables and Figures*

12. Performance of Young and Old Cows Fed Wintering Diets Containing Whole Aspen Tree Material, 82
13. Nutrient Composition of Muka Made from *Pinus sylvestris*, 83
14. Effect of NaOH Treatment on the In Vitro Digestibility of Hardwoods, 89
15. Degree of Delignification Required to Attain 60 Percent In Vitro Digestibility, 94
16. Composition and Cellulase Digestion of Various Woods Before and After SO_2 Treatment, 96
17. Lignin and Carbohydrate Content and Digestibility of Sound and Decayed Aspen and Birch Wood, 99
18. Effect of Electron Irradiation on the In Vitro Digestion of Aspen and Spruce, 100
19. Kilograms Primary Sludge per 1000 Kilograms Pulp Produced by Different Pulping Processes, 103
20. Kilograms Primary Sludge per 1000 Kilograms Pulp Produced, by Region, 103
21. Kilograms Primary Sludge Produced per 1000 Kilograms Paper Produced by Different Types of Mills, 104
22. Kilograms Primary Sludge Produced per 1000 Kilograms Paper Production, by Region, 104
23. Inorganic Content of Primary Sludges, 105
24. Sulfite Pulp and Spent Liquor Solids Production from North American Sulfite Mills, 105
25. Spent Sulfite Liquor Handling in United States Sulfite Mills, 106
26. Canadian Sulfite Pulpmills, by Process and Spent Liquor Solids Handling, 106
27. Composition and In Vitro Rumen Digestibility of Pulpmill Residues, 108
28. Composition and In Vitro Rumen Digestibility of Combined Pulpmill and Papermill Sludges, 112

ANIMAL WASTES

29. Livestock and Poultry Waste Production in the United States, 123
30. Distribution of Nitrogen in Feces and Urine from Livestock, 124
31. Mean Composition and Energy Value of Animal Wastes, 125
32. Mean Mineral Composition of Animal Wastes, 126
33. Mean Additional Mineral Composition of Animal Wastes, 127
34. Mean Amino Acid Composition of Animal Wastes, 128
35. Performance of Cattle Fed Diets Containing Dehydrated Layer Waste (DLW), 134

36. Milk Production of Cows Fed Diets Containing Dehydrated Layer Waste (DLW), 136
37. Performance of Sheep Fed Diets Containing Dehydrated Layer Waste (DLW), 137
38. Performance of Swine Fed Diets Containing Dehydrated Layer Waste (DLW), 138
39. Performance of Growing Chickens Fed Diets Containing Dehydrated Layer Waste (DLW), 140
40. Performance of Laying Hens Fed Diets Containing Dehydrated Layer Waste (DLW), 141
41. Performance of Cattle Fed Diets Containing Poultry Litter, 142
42. Results of Bacteriological Analysis of 44 Samples of Poultry Litter, 152
43. Survival of *Salmonellae* in Cattle Waste and in an Ensiled Waste-Feed Mixture, 153
44. Effect of Temperature on Survival of *Salmonellae* in an Ensiled Waste-Feed Mixture, 154
45. Microorganisms Recovered from Samples of Poultry Waste, 156
46. Effect of Temperature and Moisture on Microbial Counts of Dehydrated Poultry Waste, 157
47. Drug Residues in Broiler Litter, 161

CROP RESIDUES
48. Estimated Supply of Crop Residues, 180
49. Dry Cow Daily Weight Gain (kg) on Various Corn Residue Systems, 182
50. Pelleted and Sodium Hydroxide-Treated Cornstalks, 188
51. Performance of Steers Fed Sodium Hydroxide-Treated Corncobs, 190
52. Treated Husklage Versus Corn Silage, 190
53. Performance of Growing Calves Fed Different Chemically Treated Cobs, 192
54. Performance of Lambs Fed Wheat Straw, 195
55. Effect of Treatment of Wheat Straw and Balancing Minerals for High Sodium Intake on Rate and Efficiency of Gain of Steers, 195
56. Effect of Chemical Treatment of Wheat Straw on Lamb Dry Matter and Neutral Detergent and Acid Detergent Digestibility and Growth, 197
57. Cow-Calf Production, 200
58. Beef Production Systems with Heifers Utilizing Crop Residues, 201

xiv   Tables and Figures

AQUATIC PLANTS
59. Production of Seaweeds and Aquatic Plants, 213

APPENDIX TABLES
1. Proximate Compositon and Energy Value, 228
2. Mineral Composition, 241
3. Additional Mineral Elements, 247
4. Essential Amino Acid Composition, 250
5. Nonessential Amino Acid Composition, 253
6. Vitamin Composition of Algae, 253

Figures

NONFOOD INDUSTRIAL WASTES
1. Simplified flowsheet of production of single-cell protein, 55

FOREST RESIDUES
2. Effect of NaOH pretreatment on the in vitro digestion of straw and poplar wood, 88
3. Relationship between lignin content and in vitro digestion for NaOH pretreated hardwoods, 90
4. Relationship between level of NaOH pretreatment and in vitro digestion for quaking aspen and northern red oak, 91
5. In vitro dry-matter digestion of rations containing untreated and NaOH-treated aspen, 92
6. Relationship between in vitro digestibility and extent of delignification for kraft pulps made from four wood species, 93
7. Relationship between digestibility and extent of delignification for wood pulps, 95
8. In vivo dry-matter digestion of rations containing SO_2-treated red oak, 97
9. Relationship between in vitro digestibility and time of vibratory ball milling, 100
10. Division of the United States into six regions, based on best judgment of similar tree species, pulping processes, and end products within a region, 102

CROP RESIDUES
11. In vitro dry-matter disappearance (IVDMD) of cornstalks harvested over time, 181

12. In vitro dry-matter disappearance (IVDMD) for cobs with different combinations of sodium hydroxide (NaOH) and calcium hydroxide (CaOH), 193
13. Average daily gains of steers fed corncob ration with 0, 50, or 100 percent alfalfa hay addition, 202

AQUATIC PLANTS
14. Flow diagram of kelp dewatering process, 219
15. Harvester for water hyacinth, 220
16. Floating harvester for submerged aquatic plants, 221
17. Hyacinth control station, 222

Underutilized Resources as Animal Feedstuffs

Overview

Because conventional feedstuffs are often expensive, livestock producers have regularly been forced to seek less costly alternatives. By-products from food and beverage processing—bran, middlings, tankage, oil meals, brewers and distillers grains—represent one such class of alternatives. Some of these wastes have been used extensively as feeds, and their use has resulted in more economical livestock production. But many other potentially valuable feed sources, some of which have substantial nutritional value and are available inexpensively and in large quantities, have been underutilized. These products include food processing wastes, such as vegetable and fruit processing residues; dairy whey and tannery by-products; wastes from industrial processing and from municipalities; forest products and pulp and papermill residues; crop residues; aquatic plants; and animal waste (excreta).

There are, of course, alternative uses for some of these substances. Forest products may be used for fuel or as soil conditioners. Crop residues left in the fields are generally plowed under. Animal wastes are typically used as fertilizer, although in at least some instances the costs of hauling and spreading them are greater than the value of the plant nutrients they provide. Such uses of these materials may be economically feasible, but their use as feedstuffs will usually be more economical. Many products have no other uses.

These underutilized materials (in this report, "underutilized" materials mean those that have substantial potential value as feedstuffs but that are now used only to a limited extent) frequently present problems for disposal.

Dumping them in landfills, applying them to the land, or incinerating them is usually not feasible; in some instances it is impossible. Concentrated animal enterprises, such as feedlots, are generally located in places where areas available for land application of wastes are limited. Because these enterprises are often near municipalities, lakes, or streams, they also present potential air and water pollution problems. Similar difficulties are encountered in disposing of food and forest product processing wastes. Air pollution regulations of the Environmental Protection Agency and of other federal, state, and local monitoring agencies generally preclude incineration, and at any rate, materials containing high moisture levels are difficult to burn. The increased use of these wastes as feedstuffs would help keep such problems of disposal and pollution to a minimum.

The extent to which these substances will be used depends on a number of factors—among them availability, the attractiveness of competing alternative uses (as fuel, for instance), and the ease with which they may be used. But the most significant factors are the cost, based on nutritional value, of the processed products relative to conventional feedstuffs and their safety for both animals and humans.

Available data indicate that animal performance is generally related to nutritional value, and the nutritional value of these substances varies greatly (see Appendix Tables). Classes of underutilized substances have the following general characteristics:

• *Food processing wastes.* The generally high water content and perishability of food processing wastes require that they be used shortly after processing. Animal processing wastes are usually high in protein; plant processing wastes are usually low. The high water content of the wastes makes processing and storing difficult and expensive, although ensiling with low-moisture materials may be feasible.

• *Industrial nonfood processing wastes.* These wastes usually have high moisture content. Some, such as fermentation residues from the production of antibiotics, can be used directly, with minimum processing. Others, such as acids, alcohols, aldehydes, and esters, can be used for single-cell protein production. The possible presence of toxic organic chemicals and heavy metals represents a major obstacle to their use.

• *Forest residues.* Carbohydrates from whole-wood residues are generally resistant to ruminal cellulolytic microbes. Some chemical and physical treatments are effective to a limited extent: Generally, hardwoods are more responsive to treatment than softwoods. Wood residues are, however, usually low in protein. Some residues from the pulp and paper industries are partially delignified and are thus potentially good sources of feed for ruminants.

- *Animal wastes.* These wastes are low to fair in energy value, fair to high in crude protein, and high in minerals. The high fiber and frequently high levels of nonprotein nitrogen make them best suited for use by ruminants. Because the wastes may potentially contain pathogenic microorganisms, processing is needed. One method that appears especially feasible is ensiling with other ingredients. Quality or taste of animal products is not adversely affected by the feeding of animal wastes.
- *Crop residues.* These wastes are usually low in protein and low to fair in energy value. The energy value can be improved by certain processes; treatment with alkali or ammonia appears the most promising.
- *Aquatic plants.* Plants vary in nutritional value. Algae are high in protein and low in fiber; water hyacinths are high in fiber, indicating limited energy value, and lower in protein than algae. Satisfactory animal performance has been achieved by including these plants in animal diets. Their location and their high water content, however, present problems in harvesting and processing.

The safety of underutilized materials for use as feedstuffs must also be considered. They may contain pathogenic organisms or be contaminated with pesticides, drugs, heavy metals, or other substances toxic to livestock. Data are limited concerning chemical residues in milk and eggs from animals consuming feeds with significant levels of these residues. In meat-producing animals, appropriate withdrawal periods have been effective in preventing harmful residues in edible tissues. Sufficient data are available to ensure that many substances are effective and safe for target animals and that the resulting animal products will not compromise human health. Further research is needed, however, to ensure that feeding of some substances is safe and to satisfy the requirements of the U.S. Food and Drug Administration and other agencies regulating their use.

1
Industrial Food Processing Wastes

INTRODUCTION

As a result of changing waste-handling technology, many industrial food processing wastes now being fed to animals were once considered to be without economic value as animal feed. Other factors that have increased the interest in wastes as animal feeds include the cost of disposing of waste and increased restrictions, brought about by environmental concerns, on disposing of waste materials. The necessity of separating solid waste from liquid waste, as well as the necessity of removing suspended and dissolved substances from wastewater before discharging it, has resulted in the production of waste materials that are lower in water content and consequently more economically attractive as animal feeds. More stringent controls on the use of pesticides have also reduced the pesticide levels in food processing wastes.

QUANTITY

In general, information on quantities of industrial food processing wastes (residuals) is limited. Information on processing wastes from fruit, vegetable, and seafood processing, collected by Katsuyama et al. (1973), is presented in Tables 1 and 2. Although these data are not current, an overall view is given of waste from fruit, vegetable, and seafood processing. The major changes since 1973 would be greater utilization of wastes for animal feed and as sources of energy. There are also data on quantities of waste in the sections on specific wastes. Table 1 summarizes food processing

TABLE 1 Yearly Food Processing Industry Solid Residuals, by Product and Disposal Method

Product	Total Raw Tons Processed	Total as Solids	Total in Liquid	Feed	Other	Total By-Products	Total Residuals	Not Accounted For
Vegetables								
Asparagus	109	22	0	17		17	38	3
Bean, dry	209	5	a	2		2	6	(−1)
Bean, lima	109	9	0	9		9	17	(−3)
Bean, snap	571	61	3	58		58	118	0
Beet	245	59	5	16		16	82	19
Broccoli sprouts, cauliflower	236	19	1	82		82	100	a
Cabbage	209	58	5	5		5	69	(−1)
Carrot	254	33	2	91	a	91	127	9
Corn	2,249	81	3	1,388		1,388	1,469	38
Greens, spinach	218	7	a	22		22	30	4
Mushroom	61	29	a			0	29	(−2)
Pea	526	22	0	44	a	44	67	4
Pickle	508	36	1			0	37	(−14)
Potato, white	3,238	77	44	943		943	1,061	154
Pumpkin, squash	200	20	8	22	2	23	50	88
Tomato	6,322	345	27	109		109	472	136
Vegetable, misc.	1,106	100	47	100		100	245	a
Fruits								
Apple	952	82	a	100	79	181	263	27
Apricot	109	5	a	6	2	8	14	4
Berry	181	9	2	1		1	13	2
Cherry	172	18	1	4	a	4	24	1
Citrus	7,075	72	1	2,721	3	2,721	2,794	281
Fruit, misc.	136	23	a	7	2	9	33	3
Olive	77	2	a		9	9	10	a
Peach	998	163	13	45	40	85	263	(−17)
Pear	372	65	9	33		33	109	13
Pineapple	816	27	9	326		326	363	0
Plum, prune	24	5	1	a		0	6	1
Specialties[b]	2,268	39	22	190	15	209	272	0
Seafood								
Clam, scallop	82	11	a			0	12	59
Oyster	18	0	2		14	14	16	a
Crab	27	4	14			0	20	1
Shrimp	109	6	37	14	a	15	60	18
Salmon	109	0	32	4	2	5	36	3
Sardine	24	0	0		5	5	5	a
Tuna, misc. seafood	472	0	a	62	27	90	90	82
U.S. total	30,391	1,514	289	6,421	119	6,624	8,420	950

NOTE: All figures × 1,000 metric tons; rounded (after adding).

[a] 450 metric tons or less.
[b] Baby food, soups, ethnic foods, health food, and prepared dinners.

SOURCE: Adapted from Katsuyama et al. (1973).

industry solid residuals by product and disposal method. The weight not accounted for, when it is a positive number, represents materials probably leached into wastewater and other product shrinkage in the time between weighing the raw product and processing it. When the numbers are negative, the weight not accounted for is probably due to errors in estimating the percentage yields of the residual tons disposed of.

Table 2 shows the same data as Table 1, but on a regional basis.

PHYSICAL PROPERTIES

In general, food processing wastes contain a high percentage of water, are perishable, and must be processed rapidly. The dry matter of animal processing wastes tends to be high-protein, low-carbohydrate materials that are available throughout the year. The dry matter of fruit and vegetable processing wastes tends to be low-protein, high-carbohydrate materials that are seasonably available.

FRUIT AND NUT PROCESSING WASTES

Apple (*Malus pumila*) Processing Wastes

Several wastes from apple processing are suitable animal feeds. Apple pomace, the residual material from pressing apples for juice, contains

TABLE 2 Food Processing Industry Solid Residuals per Year, by Region

Region	Total Raw Tons Processed	Total as Solid	Total in Liquid	Feed	Other	Total By-Products	Total Residuals	Not Accounted For
New England	889	8	23	104	14	118	154	163
Mid-Atlantic	1,868	254	6	99	59	154	417	118
South Atlantic	7,546	200	62	2,436	13	2,449	2,712	299
North Central	5,342	363	25	1,156	18	1,179	1,560	56
South Central	1,106	49	23	195	6	200	272	39
Mountain	218	18	3	42		42	62	12
Northwest	3,909	136	52	1,263	12	1,279	1,460	100
Alaska	145	0	45	3	4	6	52	5
Southwest	9,351	499	55	1,186	75	1,261	1,824	118
U.S. Total	30,374	1,527	294	6,484	201	6,688	8,513	910

NOTE: All figures × 1,000 metric tons; rounded (after adding).

SOURCE: Adapted from Katsuyama et al. (1973).

pulp, peels, and cores. Between 250 and 350 kg wet pomace are formed from each ton of apples pressed for juice, or 25 to 35 percent of the fresh weight of the apple is retained in the pomace after pressing (Smock and Neubert, 1950). The residual material from canning, drying, and freezing, also called apple pomace, consists of the peels, cores, and discarded apples or pieces. Either type of apple pomace may be used for vinegar or other by-product production. Pomace is also used for livestock feed (Katsuyama et al., 1973). Leaves, stems, dirt, and some other wastes are disposed of as landfill or by field spreading. Apple pectin pulp is the residue following extraction of pectin from apple pomace. Pectin extraction is less commonly practiced than in the past because citrus pectin extraction is more competitive (Ben-Gera and Kramer, 1969).

Apple pomace from juice pressing often contains 0.5 to 1.0 percent rice hulls that are added to aid filtration (Walter et al., 1975). Dried rice hulls were added at a ratio of 1:13 on a dry-weight basis in one study (Wilson, 1971). Other filter aids include diatomaceous earth and fiber paper (Katsuyama et al., 1973).

Nutritional Value

Apple pomace and pectin pulp, wet, dried, or ensiled, are suitable feeds for ruminant animals (Smock and Neubert, 1950). Apple pomace is palatable to cattle and sheep; pectin pulp is less palatable to dairy cows than is apple pomace, and addition of molasses was suggested to increase the palatability of pectin pulps. Smock and Neubert (1950) reported that apple pomace was unsuitable for horses and of questionable value to pigs. Average digestion coefficients of wet apple pomace for ruminants are protein, 37 percent; fat, 46 percent; fiber, 65 percent; and NFE, 85 percent. Burris and Priode (1957) found apple pomace had feeding value similar to grass silage for wintering beef cattle.

Addition of rice hulls increases the fiber content and lowers the feeding value of pomace.

Processing

Fresh pomace spoils rapidly and must be used quickly or be preserved by dehydration or ensiling. Drying to about 10 percent moisture prevents spoilage and spontaneous combustion. Drying takes place in direct-fired, rotary-drum driers, and the pomace is then ground in hammer mills (Cruess, 1958). The processing may result in some heat damage to the protein.

Apple pomace is often mixed with alfalfa or corn for ensiling (Smock and Neubert, 1950). Cull apples may also be preserved as silage by mixing with 20 percent alfalfa hay. Apples ensiled alone result in a very high moisture product with considerable loss by drainage.

Citrus Processing Wastes

Wastes from the citrus industry are very well utilized as by-products, including as feed. Approximately 39 percent of the processed fruit is not used in the primary product; 97 percent of this amount is recovered as by-products (Jones, 1973). Wastewater treatment accounts for the major remaining waste problem in the citrus industry. Recent developments have centered on the use of activated sludge wastes as animal feed and are reviewed in the section, "Citrus Activated Sludge."

Hendrickson and Kesterson (1965) reviewed the processing of citrus wastes, the composition of by-products, and their utilization.

The three main by-product feeds from citrus processing are

1. Dried citrus pulp, which is formed by shedding, liming, pressing, and drying the peel, pulp, and seed residues to 8 percent moisture.

2. Citrus meal and fines, which are formed and separated during the drying process. A typical processing plant produces 85 percent citrus pulp, 14 percent citrus meal, and 1 percent fines. Citrus meal has higher density than pulp, and higher fiber, nitrogen-free extract, and ash content.

3. Citrus molasses, which is made by concentrating the press liquor from the citrus peel residue. It is usually added to the dried citrus pulp.

Most of these materials are utilized as animal feed, although citrus peel liquor and citrus activated sludge are utilized to a lesser extent. Citrus peel liquor is a by-product, similar to citrus molasses, but not as concentrated (S. Reeder, SunKist Growers, Inc., Ontario, Calif., 1979, personal communication). The citrus peel liquor studied by Lofgreen and Prokop (1979) had a dry matter content of 47.3 percent.

In feeding trials with growing beef cattle, citrus peel liquor was found to have a net energy for maintenance (NE_m) of 2.24 Mcal/kg and for gain (NE_g) 1.48 Mcal/kg on a dry-matter basis (Lofgreen and Prokop, 1979). Citrus peel liquor fed with cane molasses contained higher net energy values than either one alone. NE_m and NE_g are 1.97 and 1.32 Mcal/kg, respectively, for citrus molasses, on a dry-matter basis (National Research Council, 1976).

Citrus activated sludge is produced by treating the liquid wastes from citrus processing plants. Sludge recovery and the nutritional value for poultry have been studied (Damron et al., 1974; Jones et al., 1975). Dehydrated sludge was found to be profitable. Dried sludge was acceptable as a poultry feed for up to 7.5 percent of the diet (Damron et al., 1974). Its inclusion in the diet reduced the amounts of yellow corn, soybean meal, and phosphorus required. However, amounts of fat and methionine had to be increased to maintain energy and sulfur amino acid levels, respectively. The dried sludge also improved skin and egg yolk color, and no off flavors were detected in egg yolks or albumin. Higher levels of dried sludge in the diet had detrimental effects.

Peach (*Prunus persica*) Processing Wastes

Peaches are graded and the percent of cull fruit determined before shipment of the entire lot from orchard to processors. Batches of cull fruit may need to be sorted in order to be accepted, or the grower may choose to dump the fruit in the orchard to avoid sorting costs. At the processing plant, cull and undersized fruit are removed.

The total peach processing residual is 26 percent of raw fruit. The disposal methods used are land disposal, 63 percent; liquid waste, 5 percent; feed, 17 percent; and other by-products, 15 percent (Katsuyama et al., 1973).

Usually peeling is done by the use of lye solution and washing. Dry caustic (lye) peeling methods have been investigated but are less commonly used. The waste from lye peeling is either soluble or in very fine particles and is highly alkaline. Dry caustic peeling waste sludge has the following properties: 9 to 10 percent solids content, pH of 13 to 14, dark brown color, and applesauce-like texture (Smith, 1976). Culls and pieces are removed at various stages of the processing operation. Dry wastes, including trash and cull fruit, are usually used for landfill or soil application; some fruit is used as feed or for alcohol production. Screened solids are removed from the wastewater and disposed of as above (Katsuyama et al., 1973). Peach peeling slurry from wet-peel methods may be recovered from the wastewater by shaker screen sedimentation treatment, or it may be discharged into the wastewater stream. Dry peeling wastes are kept out of the waste stream (Gray and Hart, 1972).

The factors that hinder the use of peach wastes as animal feed include (1) costs of transportation of the high-moisture wastes from processing plants to livestock producing areas; (2) the high sugar content of peaches, which makes drying difficult (and a suitable drying technology has not been developed); and (3) the short canning season.

Little information is available on the feed value of peach wastes.

Pear (*Pyrus communis*) Processing Wastes

In the past, pear waste was processed into two feed products, pear pulp (or pomace) and pear molasses. This is no longer done on a commercial scale. The 1979 production of pears in the United States was 781,553 metric tons with 502,680 metric tons or 64 percent processed (U.S. Department of Agriculture, 1980). According to Katsuyama et al. (1973) about 29 percent of the raw pear tonnage was residual and 30 percent of the residual was recovered as feed. In 1950, when pear waste processing was being developed, 40 percent of the raw tonnage was residual (Graham et al., 1952). From one ton of waste, 55 kg of pear pulp (8 percent moisture) and 135 kg of molasses (50 percent sugar content) can be obtained (Brown et al., 1950).

Pears are peeled both mechanically and chemically (lye peeling), but mechanical peeling predominates. Present methods of disposing of pear wastes include landfill and field spreading, with some of the solid wastes (peels, cores, and screened solids) being used as feed. Because of high transportation costs these wastes are fed fresh by local farmers (G. York, University of California, Davis, 1979, personal communication). Other methods of waste recovery have been considered and some implemented. These include fermentation to produce alcohol, methane production, and edible juice recovery.

Nutritional Value

According to Guilbert and Weir (1951), pear pulp and pear molasses are both very palatable to ruminants. The pear pulp studied contained a small amount of other fruit wastes. In feeding trials, steers were fed a fattening diet with pear pulp forming 25 percent of the concentrate, replacing molasses and dried beet pulp. Pear pulp had a value of 70 to 75 percent of molasses and dried beet pulp, and pear molasses had a value of 115 to 120 percent of cane molasses. Pear molasses had a lower nitrogen and ash content and higher total digestible nutrients than cane molasses. Pear molasses was more palatable than beet molasses—so palatable that it could not be fed as free choice but instead had to be mixed with other feeds. Feeding trials with sheep were also conducted (Guilbert and Weir, 1951). The total digestible nutrients, dry-matter basis, for pear pulp and pear molasses were 60 and 86 percent, respectively.

Processing

According to Brown et al. (1950) pear wastes processed for feed and containing some other wastes had an average solids content of 13.8 percent. The wastes included variable quantities of peach, tomato, and grape wastes. The variability of the composition of the waste made processing more difficult. Pear waste could not be dehydrated directly and the slimy character of the waste made pressing difficult. To overcome these problems, a process was developed, and later used commercially, consisting of grinding the waste in a hammer mill, liming, aging for a short time, and pressing. The press juice was concentrated to form molasses and the press cake shredded to form pear pulp. Because of the low fiber content of the waste, a certain amount of dried pulp or other material was added to aid pressing.

Fruit Canneries' Activated Sludge

In several food processing industries, wastewater is treated by activated sludge methods, and the feeding value of some form of activated sludge was studied. Since many fruit and vegetable processors process several fruits or vegetables at one plant and could use activated sludge treatment, this topic is covered separately.

Waste activated sludge treatment was evaluated at a plant processing apples, pears, peaches, plums, crabapples, and cherries (Esvett, 1976). The principal costs of the system were for additions of nitrogen and phosphorus because the wastes were low in these, and energy costs for aeration, sludge circulation, and sludge disposal. Sludge at the plant is currently thickened and disposed of by application to agricultural land.

Physical Characteristics

The sludge is highly viscous and can be moved by a bucket-type loader or belt conveyor (Esvett, 1976). When the sludge was stored for 10 months in a 208-liter drum, deterioration and production of ammonia were evident.

Nutritional Value

The composition of the sludge, dry-matter basis, was 39.1 percent crude protein, 3.2 percent crude fiber, 0.8 percent ether extract, 11.64 percent ash, 1.08 percent calcium, and 1.28 percent phosphorus (Esvett, 1976).

In a digestibility study, biological solids (or concentrated activated sludge) were fed to steers as 2.3, 4.5, and 9.2 percent of the diet, on a dry-matter basis. The digestibility of the diets was not affected by inclusion of biological solids at the 5 percent level or less. A second study was done using biological solids in the finishing diet of steers at 2.3, 4.6, and 8.9 percent biological solids. Feed efficiency was not significantly different from the control.

Winery Wastes

Wine-making is a large industry in California, with lesser quantities being produced in New York and other states. In 1979, 2,461,324 metric tons of grapes were used for wine-making in the United States. Of this amount 96.5 percent was processed in California, 2.5 percent in New York, and the balance in other states (U.S. Department of Agriculture, 1980). Some wines are made from apples, pears, and other fruits; information on the quantity of nongrape wines produced is not available. The major wastes from wineries are stems, pomace, lees, stillage, and cleanup washwater. If distilled wines are made, the pomace becomes part of the stillage (also called still slops or still bottoms). The stillage is the residue remaining after all alcohol has been distilled. The lees are the sediments that settle out during the storage and aging of wine and that are removed during racking. Winery operations are seasonal; crushing and fermentation occurs from August to November, while distillation may continue throughout the year.

The stems (approximately 5 percent of the original grape material) are usually disposed of by field spreading or burning (Stokes, 1967). Use of dried pomace as a feed declined to negligible amounts in 1966 (Amerine et al., 1972), but recently large amounts of pomace are being dehydrated for feed in California. The quantity of pomace produced has been estimated as 10 to 20 percent of the original grape weight (Ben-Gera and Kramer, 1969). Another study estimated a yield of 10 percent pomace, containing 45 percent solids (Pattee, 1947). Amerine et al. (1972) estimated 12 percent pomace yield. The solids content of the pressed pomace may be 30 to 35 percent (J. Cooke, University of California, Davis, 1976, personal communication). Annual U.S. production of grape pomace was estimated at 45.4 to 72.6 million dry kg, and production of combined apple and pear pomace was estimated at 7.3 to 9.1 million dry kg (Prokop, 1979). In California, approximately 18.2 million dry kg are used for feed (G. Cooke, University of California, Davis, 1979, personal communication).

Nutritional Value

Only a few studies on the feeding value of grape pomace have been conducted, but pomace has been fed to ruminant animals in the United States and other countries for many years. The French use pomace as feed and consider it similar in quality to good hay (Amerine et al., 1972). Because of differences in wine production in the two countries, pomace from French wineries may have higher feed value. According to Amerine et al. (1972), the grape stems contain appreciable amounts of fermentable carbohydrate. Grape pomace has a high fiber content and lower feed value than other fruit-wine pomaces. The seed content of wet pomace ranged from 20 to 30 percent (Amerine et al., 1972). The seeds have a fibrous hull that decreases the feeding value of the pomace. If the seeds were hulled and pressed for oil extraction; the presscake would be valuable as animal feed. The pomaces are high in moisture and generally rather fibrous.

Protein and energy are poorly digested, the digestibility of protein being 16 percent and the TDN values ranging from 24 to 30 percent, air-dry basis. These values were obtained in a study in which grape pomace was fed at a level higher than 50 percent of the diet, with alfalfa hay in one trial and as 100 percent of the diet in another (Folger, 1940). Prokop (1979), in a more recent study, fed grape pomace at 20 percent of the diet. Net energy values obtained with beef cattle on finishing diets were: NE_m, 0.75 Mcal/kg, and NE_g, 0.41 Mcal/kg on a dry-matter basis. In the same study, a combination of apple, pear, and grape pomaces and apple pomace were also tested at the 20 percent level. On a dry-matter basis, the value of apple winery pomace was comparable to beet pulp. The combination was similar in nutritive value to dried alfalfa pellets (20 percent protein). The value of pear pomace was estimated at 89 percent of the value of the apple pomace (Prokop, 1979).

In a recent study in Cyprus, Hadjipanayiotou and Louca (1979) fed dried grape pomace (also called grape marc) as 15 and 30 percent of calf fattening diets. Urea was added to compensate for the low digestibility of grape pomace protein. The composition of the grape pomace was very similar to that reported by Prokop (1979). On a dry-matter basis, crude protein was 12.3 percent and was 19.5 percent digestible. The digestion coefficient for dry matter was 28.4 percent. Metabolizable energy was 1450 kcal/kg, dry-matter basis—approximately half that of barley. The feed intake and gain of the calves were not significantly different when grape pomace formed 0, 15, or 30 percent of the diet. However, there was a significant difference in feed efficiency between the 0 and 30 percent levels. The dressing percentages of the calves fed 30 percent grape pomace tended to be lower.

Most of the pear and apple winery pomaces are being used as feed. Some is being dried for this purpose, but most is being fed wet, (M. J. Prokop, University of California, El Centro, 1979, personal communication).

Processing

Wet grape pomace stores well in compacted piles; only the outer layer deteriorates (Stokes, 1967). Large quantities can be dried in rotary drum driers and ground in hammer mills. Small quantities of apple and pear pomaces may be combined with the grape pomace for drying. The non-grape pomaces are more difficult to dry, and combining with grape pomace makes drying easier. The pomace not being used as feed is spread on land.

Cacao (*Theobroma cacao*) Processing Wastes

Commercial cacao beans are imported into the United States for chocolate and cocoa manufacture. Imports of cacao beans into the United States in 1979 were 167,881 metric tons (U.S. Department of Agriculture, 1980).

The two major uses of cacao bean shells are as animal feed and nursery mulch. Calculated yields of shell available for by-product recovery ranged from 8 to 12 percent of the commercial beans, depending on the efficiency of winnowing, the quantity of shell-nib mixtures, and the variety of bean (Chatt, 1953).

Physical Characteristics

The shells are brittle and, depending on the processing equipment, may be broken into fine pieces, but they are not ground intentionally at this stage. Feed manufacturers may grind the shells prior to incorporating them into feed mixtures. In California more shells are used as feed than in other areas of the country because cacao shells are not used as extensively for mulch in California, other mulches being preferred.

Nutritional Value

The fat content of the shells is about 3 to 3.5 percent and varies with the amount of fat transferred during roasting and the quantity of nib present (the nib is the cotyledon of the cacao from which various chocolate and cocoa products are made) (Chatt, 1953). The ash content varies from 5.5 percent (Chatt, 1953) to 10.8 percent (Morrison, 1956) dry-matter basis. The crude protein content is 13.5 to 16 percent but has low digestibility (Chatt, 1953; Morrison, 1956). Crude fiber is 16 to 20 percent (Chatt, 1953). The shells also contain considerable vitamin D (Morrison, 1956).

Fruit Pits, Fruit Pit Kernels, Nut Hulls, and Nut Shells

Total U.S. nut shell and fruit pit wastes have been estimated at over 453.6 million kg annually (Mantell, 1975). Nut shells and fruit pits are burned as fuel, made into charcoal, or used as landfill. Some of the charcoal made from nut shells and fruit pits is used in animal feeds. Shells from peanuts, almonds, pecans, and filberts can be used as mulch, but must compete with many other waste products. Almond and peanut shells can be used as poultry litter. In general, fruit pits (except kernels) and nut shells have not been found to be useful livestock feed materials. They are now being burned as energy sources at fruit and nut processing plants.

VEGETABLE PROCESSING WASTES

Potato (*Solanum tuberosum*) Processing Wastes

The potato is processed into many different forms, usually at separate plants. In 1978, 7.95 million metric tons were processed (U.S. Department of Agriculture, 1980). The products include dehydrated potatoes (19.1 percent); chips and shoestrings (22.1 percent); frozen french fries (45.5 percent); other frozen products (8.9 percent); canned potatoes and other canned products, such as soups, hash, and stews (2.7 percent); and starch and flour (1.7 percent). Potato starch production results in quite different waste products and will be covered separately.

In 1971 it was estimated that 33 kg of solid waste were generated per 100 kg of potatoes processed (Jones, 1973). Of this, 29 kg (88 percent) of solid potato waste is turned into by-products, mostly feed. The other 12 percent is handled as solid waste, and some of this is recoverable as feed.

About 5 percent of the potato crop was fed to animals in the United States in 1966 (Ben-Gera and Kramer, 1969). In 1971 only about 3 percent of the fresh crop was used as feed (Hogan and Highlands, 1975).

Physical Characteristics

Wastes from potato processing include (1) the large solids that are removed by screening (peels, culls, trimmings), (2) the smaller solids that are removed by primary treatment, and (3) less easily separable solids and dissolved solids that may be treated by secondary treatments or used as substrate for yeast or fungal growth.

The large solids are known as screening or screen solids. The solids removed by primary treatment may be referred to as primary solids, sludge, slurry, or filtered potato sludge. The primary solids include pulp, the more finely divided pieces from which the solubles have been leached.

The effluent from primary treatment contains solubilized starch, proteins, amino acids, and sugars (Pailthorp et al., 1975). This effluent may receive secondary treatment, the resultant sludge being called waste biological solids, bacterial mass, or waste activated sludge.

Potato by-product meal is produced from the screenings and other wastes of products for human consumption. It also may include cull potatoes and potato pulp from starch manufacture (Dickey et al., 1971). Limestone is added to mixed wastes to aid drying and the mixture is dehydrated. Each waste material is metered into the mixture so that the resultant by-product is fairly uniform.

Alkaline potato peels from dry caustic peeling have a pH of about 12, and solids content of 12 to 15 percent. The peelings can be used to neutralize straw that has been steam treated under pressure. This is a patented process (R. P. Graham, U.S. Patent 3,692,530, September 19, 1972).

Potatoes have been sun-dried on abandoned concrete airstrips in California (Hogan and Highlands, 1975) and have been "freeze-dried" in the Red River valley and in Colorado by spreading on pastures in the winter. Cattle are allowed to feed on the partially dried potatoes in the field.

The Symba-yeast process is a method of deriving a food or feed product from waste effluents including primary solids (Skogman, 1976). The dry matter content of the wastewater should be over 2 percent. This can be increased by wastewater recirculation or adding solid waste. The production season is fairly long.

In the *Aspergillus niger* protein process, starchy material is homogenized to increase surface area; inorganic nitrogen and other nutrients are added; and the material is heated, inoculated with fungus, and aerobically cultured for 24 to 36 hours. The product is then sterilized again and filtered (Gillies, 1978) before being used in the preparation of feeds.

About 200 to 225 kg of pulp solids are produced per ton of starch produced (Treadway, 1975). Protein water, also known as fruit water and starch washwater, usually contains 1 to 3 percent solids (Treadway, 1975). About 225 to 325 kg of protein water solids are produced per ton of potato starch.

Nutritional Value

Primary solids and slurry are usually dilute, containing 4 to 7 percent solids. These may be concentrated by belt-type vacuum filters (preferred method) or by centrifuges to 15 to 18 percent solids (Pailthorp et al., 1975). When fed to calves as 25 to 50 percent of concentrate mixture, primary solids were equivalent to ground barley (Hogan and Highlands, 1975). The dry matter was 73.5 percent digestible and contained 1.498 kcal ME/kg when fed at 19.2 percent of dry matter of the diet and 1.259 kcal ME/kg when fed at 37.5 percent of dry matter of the finishing diet of steers. It can be fed at 40 to 50 percent of dry matter of a steer diet before a reduction in daily gain occurs (Hogan and Highlands, 1975).

Primary solids from caustic peeling plants have a high sodium hydroxide content and cannot be fed at high levels (Grames and Kueneman, 1969). The sludge can be stored prior to filtration, and fermentation will cause the pH to drop and increase filterability. The product can then be fed in combination with other feeds, as is presently being done at some plants. Some plants run their own feedlots to dispose of the sludge. The price obtained for primary sludge is sufficient to pay for processing. The alkaline peel-straw product has a pleasant molasses-like odor. The solids content is about 20 percent, and represents a combination of 1.9 parts steamed straw and 1 part potato peels, on a dry-matter basis. The product has a pH of about 7 and a total-soluble-after-enzyme in vitro digestibility of 80 to 85 percent.

The product from the Symba-yeast process has been tested in feeding trials with calves as 40 percent replacement for milk proteins (Skogman, 1976). The limit is about 25 percent of poultry diets. It has also been tested as a portion of the diet of rats, pigs, mink, and house pets. The yeast product is low in nucleic acids, 4.25 percent dry-matter basis, and has high B-vitamin content. There is no difference in the composition of the yeast made from different waste materials. The ash content in yeast from lye peel water may be higher.

The *Aspergillus niger* protein process fungal product contains 37 percent protein (Rogers and Coleman, 1978).

Fresh potatoes are fed to ruminants and horses. Swine and other monogastrics should have potatoes cooked to increase digestibility (Hogan and Highlands, 1975). Fresh potatoes can be stored by ensiling with a minimum of 20 percent dry corn stover or hay. Cooked potatoes can be ensiled without roughage (Hogan and Highlands, 1975). Potato silage that includes 20 percent corn fodder or alfalfa hay is approximately equal in value to corn silage for cattle and sheep. Raw potato has lower apparent digestibility of nitrogen and lower efficiency of nitrogen utilization in pigs. Chymotrypsin inhibitor activity is high in raw potato and absent when cooked (Whittemore et al., 1975).

Cooked potatoes are equivalent to corn in nutritive value for fattening pigs; however, the high moisture content reduces feed intake. Cooked potato flakes have been studied in feeding trials with chicks and pigs (D'Mello and Whittemore, 1975). The ME energy value of cooked potato flakes in chicks is 3.26 Mcal/kg, dry-matter basis. Growth rates and efficiency of feed conversion were lower in chicks fed cooked potato flakes than in those fed glucose or corn. Up to 20 percent cooked potato flakes in pelleted diets allowed satisfactory performance; 30 to 40 percent had adverse effects on growth and efficiency. Feed intake was not depressed by potato flakes. The protein quality of cooked potato flakes was comparable to that of ground corn.

When cooked potato flakes were fed to piglets, better results were obtained than when fed to poultry (Hillyer and Whittemore, 1975). The digestibility of energy was 96 percent and of nitrogen, 89 percent. Cooked potato flakes contained a DE of 3.94 Mcal/kg, dry-matter basis. Diets containing 50 to 58 percent cooked potato flakes were equal in nutritive value to those containing corn. The potato flakes contain slightly less sulphur amino acids and slightly more lysine than cereals.

The digestibilities of dry matter and nitrogen by calves decline when cooked potato flour replaces spray-dried whey in liquid diets (Hinks et al., 1975).

Wet pulp is commonly fed fresh to cattle in Europe. Pulp silage is very palatable to cattle (Brugman and Dickey, 1955). Dried potato pulp is high in sugar, low in fat, moderate in fiber, and low in vitamins. Manufacturers often add 2 to 6 percent molasses to the pulp. Digestible crude protein is 6 percent and TDN is 79 percent for cattle. It is equivalent to yellow hominy feed at 22.5 percent of the grain mixture. Dried potato pulp may have too high a fiber content for hogs but could be used as a portion of grain. It is assumed that potato pulp would have a value for hogs similar to that for cattle and other ruminants but little information is available. The fiber content of dried potato pulp is too high for poultry, but the pulp may be used to limit energy intake for replacement stock (Brugman and Dickey, 1961).

Dried pulp can be used to improve grass silage and as a preservative in silage. It has excellent water-holding capacity, 210 kg of water per 100 kg of pulp.

About 60 percent of the total solids of protein waste water are nitrogenous compounds, two-thirds of which is free amino acids, and one-third protein (Treadway, 1975). The nonnitrogenous solids include sugars, organic acids, inorganic salts, and a few other minor constituents.

A feeding study was done with chicks using whole fruit water concentrate meal that contained 50 percent protein, dry-matter basis (Rosenau et al., 1977). This product slowed chick growth rates if used above the 6 percent level. This was thought to be caused by lysine loss due to Maillard browning. The presence of trypsin inhibitor in potato products requires that the product be heat-treated in order to inactivate the inhibitor.

Processing Methods

The waste effluent from primary treatment can be treated by numerous types of secondary treatments (Guttormsen and Carlson, 1970; Richter et al., 1973). Most potato processors have primary treatment but fewer have secondary treatment, although increasing numbers are applying secondary treatment as pollution standards are raised. More study has been done on starch-plant soluble waste recovery than on potato processing plant soluble wastes.

The yield of biological solids from activated sludge process is large; about one unit of biological solids was produced per unit of biological oxidation demand (BOD) removed (Richter et al., 1973).

Wet pulp is difficult to dewater by pressing because of its slippery texture and high water holding capacity. Treating the wet pulp with 0.3 to 0.5 percent $Ca(OH)_2$ for 20 minutes prior to pressing improves water removal.

In Germany it was reported that pulp is pressed to reduce water content to 70 to 75 percent by using a double drum centrifugal press. This pulp was then dried and used as cattle feed (Hogan and Highlands, 1975). In Maine pulp was pressed or centrifuged to 79.5 to 83 percent water (Brugman and Dickey, 1955). Pulp can be dried in a steam tube drier or direct-fired rotary drier. Pulp is commercially dried in Maine by means of an add-back process and is not pressed prior to drying (Hogan and Highlands, 1975).

Wet pulp, if stored, ferments rapidly. After 8 days in storage, pH drops to 3.7 (Brugman and Dickey, 1955). It will drop further over time to 3.5 and store well for extended periods. Silage may be enriched with protein that has been extracted from fruit water by precipitation with organic acids (Ben-Gera and Kramer, 1969).

Studies have been concentrated on the recovery of the various constituents of protein water from production of potato starch (Osten, 1976). Proteins are in fairly low concentrations in the water, usually 1.1 percent solids. This can be increased to 3.1 percent solids by using a starch recovery process that uses less water. Efforts have been made to decrease the amount of water used by changing the starch recovery process, and efforts have also been made in numerous methods of protein extraction and/or protein water concentrations.

Rosenau et al. (1976) described a starch recovery process using greatly reduced amounts of water, yielding a protein water with 4 percent solids and 50 percent protein, dry basis. Rosenau et al. (1977) mentioned a process yielding a juice stream with 5 percent solids.

Four methods of constituent recovery of protein are evaporation, ultrafiltration, coagulation of protein by steam injection heating and concentration of filtrate, and the five-step method of constituent recovery. Evaporation was used to concentrate fruit water solids (Stabile et al., 1971). The protein water is evaporated in a triple-effect evaporator to 60 percent solids. This slurry could be mixed with dried potato pulp for use as feed. A yield of 24 tons per day of mixed feed is postulated from a 27 ton-per-day starch plant.

Della Monica et al. (1975) studied the stability of protein water slurry. It was found that the product would require final drying. Whole juice concentration and spray or drum drying yields 50 percent crude protein meal. Sufficient heating is necessary to inactivate the trypsin inhibitor.

Rapid heating of fruit water by steam injection and acid precipitation can be used to coagulate about 35 percent of the protein (Rosenau et al., 1977). The protein product does not have antitrypsin activity. The end product contains 8 percent moisture and 75 to 80 percent protein on a dry-matter basis. This process has a high energy requirement (Osten, 1976). The deproteinated juice is then concentrated to 70 percent solids. Because of lysine destruction, Rosenau et al. (1977) suggested mixing it with pulp and feeding it to ruminants. By-product yields were estimated at 1.1 kg of protein meal (10 percent moisture) and 8.9 kg of pulp (10 percent moisture) from 100 kg of potatoes processed for starch. Starch production from 100 kg of potatoes would be 12.2 kg of starch (18 percent moisture). The juice may be processed further by heating to 60°C in a plate-type regenerative heater and then to 121°C by steam injection. This coagulates about 35 percent of the crude juice protein so that it can be removed as a 20 percent solids (70 percent protein, dry basis) sludge by a nozzle-type disk centrifuge and dried in a spray drier. The remaining deproteinated juice (43 percent crude protein, dry basis) is concentrated by reverse osmosis to 10 percent solids and by multiple-stage evaporation to 65 percent solids. It is stable in this condition and could be useful as a molasses substitute for ruminant feeding (Rosenau et al., 1978).

Sweet Potato (*Ipomoea batatus*) Processing Wastes

Sweet potatoes are grown mostly for human consumption today, although they have been grown for animal feed. They are sold fresh, canned, or frozen.

The production of waste from conventional sweet potato canning can be divided as follows (Colston and Smallwood, 1972): 5 percent of the weight of incoming sweet potatoes is dirt, which is removed by screening and washing and is disposed of as landfill; 9.5 percent of original potatoes is culled (cull potatoes can be used as cattle feed); 40 percent of the original becomes the canned product; 45.5 percent of original is solid waste.

According to Smallwood et al. (1974), snips and trimmings are removed by 10-mesh screen, are handled as solid waste, and account for 20.5 percent of the original potatoes. These have been disposed of as landfill but could be removed as feed. Twenty-five percent of the original potato appears in the waste stream after screening. The waste effluent has a pH of 9.5 to 11.5. A large amount of the 25 percent original could be recovered by a two-stage screening system.

Nutritional Value

Sweet potato meal was equivalent to corn when fed to cattle at less than 50 percent of the grain portion of the diet (Bond and Putnam, 1967). If sweet potato meal is used to replace all of the grain in the diet, lower performance results. As a large portion of the diet, sweet potato meal is less palatable than grain and can be laxative (Morrison, 1956).

If cottonseed meal is added to sweet potato meal to bring its protein content up to the level in corn meal, the sweet potato meal, in general, is equivalent to 90 percent of corn meal when fattening steers (Edmond and Ammerman, 1971). Studies with lactating cows also showed sweet potato meal to have 90 percent of the value of corn meal.

Feeding sweet potato meal to cattle and sheep has produced satisfactory results (Morrison, 1956). When used at levels up to 50 percent of the grain mixture, this feed has been equal or nearly equal to corn for dairy cows. When it replaced up to 50 percent of the grain for fattening cattle, it has been found to be worth 95 to 100 percent of the value of corn.

Blanched sweet potato meal may be equivalent to corn meal for pigs, but not enough research has been done to draw definite conclusions. Studies have shown that sweet potatoes are less economical to grow for feed than corn; however, culls and processing wastes, when dehydrated or dried, can be used as feed (Edmond and Ammerman, 1971).

Bond and Putnam (1967) studied the feeding of dehydrated sweet potato trimmings to steers. A diet with 51 percent trimmings was compared to a similar diet with 51 percent cracked corn. Sweet potato trimmings had a value equivalent to 80 percent that of corn and were not equivalent to dried whole sweet potato meal, which had a value nearly equal to that of corn. The trimmings were coarse and hard, lowering the feed intake. Grinding improved feed intake and subsequently improved the weight gain of the steers. Trimmings should be ground for best results. The steers fed trimmings had higher carcass grades but lower dressing percentage and less marbling. The digestibilities of dry matter and crude protein of the trimmings were lower than for corn (Bond and Putnam, 1967).

Processing Methods

Solid wastes from canning, cull sweet potatoes, snips, and trimmings are easily recovered (Smallwood et al., 1974). They are handled as a group and made into dry sweet potato meal, or the snips and trimmings are handled separately and made into a product called sweet potato trimmings.

Other wastes are carried as liquid waste from which solids can be recovered by screening. The screened solids include caustic peelings, abrasive peelings, and solids from washings and spillage. Screened solids may also contain snips and trimmings if they have not been handled separately. Trimmings and screened wastes are less well utilized than sweet potato meal. It was found that the cost of drum drying conventional waste was too high (Smallwood et al., 1974). Wastes from the "dry peelers" can be mixed with trim wastes and fermented to reduce the high pH. It was also noted that the cost of drying sludge from sweet potatoes was higher than for drying sludge from white potatoes. In sweet potato processing, peel losses are higher, the dry-matter content of the potatoes is higher, and more lye is used.

Tomato (*Lycopersicon esculentum*) Processing Wastes

Tomatoes are used for many canned or heat-processed products. The quantity of tomatoes processed has increased greatly in the past few decades, tripling in the 25-year period prior to 1975 (Brandt et al., 1978). Fifteen percent of processed tomatoes are used for canned whole tomatoes, a process which requires peeling, and the remaining 85 percent are used for pulped products (Schultz et al., 1976).

Tomato processing wastes can be divided into three categories according to the type of by-product that can be recovered. Cull tomato feed is made from cull tomatoes processed into tomato pulp or tomato pomace. A second category is peel residue, a waste product from canning whole tomatoes; about 12 percent of the original tomato is removed as peel and adhering pulp. Peeling residue in California in 1974 was estimated to be 170,000 tons (Schultz et al., 1976). The third category is tomato pomace, the residue from the manufacture of juice, paste, puree, sauce, and catsup. Pomace contains the skin, seeds, fiber, and some adhering pulp.

Physical Characteristics

The peel residue from caustic rubber-disc peeling has a pH of 13 to 14; a solids content of 7 to 8 percent; a bright red color; and an easy-flowing texture like a tomato puree with many skins. The pH of the wastewater from dry caustic peeling is 6.2 compared to 9.4 for wet caustic peeling (Smith, 1976).

Nutritional Value

The ash and nitrogen-free extract levels are higher in cull tomato feed than in tomato pomace, and fiber is lower (Ammerman et al.,1965). Cull tomato feed has been used to replace citrus pulp as 10 to 30 percent of the concentrate in steer diets and has produced similar results (Ammerman et al., 1963). It has been studied in the dry form. In another study, cull tomato feed has been fed as 70 and 100 percent of steer diets. Digestibility decreased slightly when cull tomato feed was fed alone. The average TDN for the two diets was 64.8 percent, dry-matter basis. No problems were encountered, although the steers seemed to be getting insufficient fiber from cull feed alone.

In a study with lambs the protein of cull tomato feed at 33 percent of the diet was less digestible and of lower biological value than that of soybean meal (Ammerman et al., 1963). On a dry-matter basis, cull tomato feed had 24 percent crude protein and 12.2 percent apparent digestible protein. In steers the apparent digestible protein was 13.5 percent. Tomato cull feed satisfactorily replaced alfalfa meal as 3 percent of the diet for poultry (Ammerman et al., 1965). Carotenoid pigmentation in skins and shanks, however, was reduced.

A level of 12 to 15 percent tomato pomace in cattle feed produced good results (Esselen and Fellers, 1939). Morrison (1956) reported that tests had been conducted with dairy cows in which pomace was included as 15 percent of the concentrate mixture with satisfactory results. Tomato pomace is palatable to hogs, cows, and chicks as 10 to 15 percent of the diet, but fed at higher levels, bitterness may make pomace unpalatable. Dried tomato pomace has a content of about 25 percent crude protein, 15 percent ether extract, 22 percent crude fiber, and 3 percent ash, dry-matter basis. It is a good source of some B-vitamins and a fair source of vitamin A (Esselen and Fellers, 1939).

More recently, feeding trials have been conducted with sheep (Hinman et al., 1978). Tomato pomace (wet) was fed at levels up to 77.5 percent of the diets, dry-matter basis. After the sheep became used to the pomace they consumed it readily. The crude protein level of all diets was about 20 percent, dry basis. The pomace had slightly lower digestion coefficients for dry matter, organic matter, and crude fiber than alfalfa, but had slightly higher coefficients for cellulose, nitrogen-free extract, and ether extract in diets with large amounts of pomace. In a second trial, lower digestibility of protein resulted when dried pomace was fed. This may have been caused by heat damage of the protein during drying. One problem found with pomace was the variability of dry-matter content; it ranged from 11.9 to 27.5 percent.

Some dried, bagged pomace is used in pet food (Katsuyama et al., 1973) and for fur-bearing animals (Schultz et al.,1976), although the portion recovered and used as animal feed in the United States is small. Tomato pomace has been used in dog foods and fox and mink diets (dry pomace at 5 percent of wet ration) to prevent diarrhea (McCoy and Smith, 1940).

Tomato pomace from tomato juice production was fed to rats to study the protein quality (Esselen and Fellers, 1939). The dry tomato pomace contained 20.9 percent crude protein, 17 percent ether extract, 12 percent crude fiber, 46.4 percent nitrogen-free extract and 3.6 percent ash, on a dry-matter basis. The protein efficiency ratio (PER) of dry tomato pomace unsupplemented with amino acids was 2.18, and net protein utilization (NPU) was 0.55. Tomato seed presscake has been studied as a feed ingredient. Rats grew at a normal rate using presscake as the sole protein source. Amino acid studies have been conducted on tomato seed presscake, tomato seeds, and tomato skins (Tsatsaronis and Boskou, 1975).

Other Vegetables

In general, when vegetables are harvested, as much of the human nonfood plant material as possible, including leaves, stalks, stems, and pods, is left in the field. Some of these are unavoidably brought into the packing or processing plant and become part of the waste residual, along with cull, broken, or unacceptable vegetables. Also included in vegetable wastes are stems from leaf vegetables, cobs and husks from corn (*Zea mays*), and cores from cabbage (*Brassica oleracea capitata*). Most vegetable wastes are disposed of and not dried because the moisture content is too high to enable the wastes to be burned.

The main value of leaf meals appears to be as a pigment and vitamin source for poultry. Ben-Gera and Kramer (1969) mentioned the use of dried celery (*Apium graveolens*) tops as cattle feed. Of the wastes discussed only pimento (*Pimenta officinalis*) wastes and possibly artichoke (*Cynara scolymus*) wastes are currently being dried (Livingston et al., 1976).

The following is a review of information on dehydrated vegetable wastes (Katsuyama et al., 1973). Although many of these wastes are fed to livestock, information is not readily available on their feeding value.

Asparagus (*Asparagus officinalis*) waste includes butt ends and broken spears. Most wastes are carried by water, and the screened wastes are used for animal feed. Asparagus waste was dehydrated, ground in a hammermill, and fed to sheep and dairy cows (Folger, 1940). The composition of the meal was 17.2 percent crude protein, 39 percent nitrogen-free extract, 1.1 percent ether extract, and 35.1 crude fiber, on a dry-matter basis. Folger found that the dried waste could be substituted for fair-quality roughage. It was palatable to sheep but less palatable to dairy cows. TDN was 51.7 percent, on a dry-matter basis.

Artichoke (*Cynara scolymus*) waste was dehydrated in a pilot plant study (Livingston et al., 1976). The solids content of fresh trimmings during the harvest peak was 16 percent. The protein content of unpressed dried meal ranged from 17.4 to 23.1 percent and that of the pressed dried meal from 14.2 to 21 percent. Pressing increased crude fiber content of the meal. The carotene and xanthophyll contents were too low for use as a poultry pigmentation supplement but would be adequate for most animal feeds. The amino acid and mineral compositions were also studied.

Lima beans (*Phaseolus limensis*) are threshed either by mobile field viners or by stationary viners. Vines and pods from field viners are plowed under, but at stationary viners these may be collected for feed use. Quantities of vines and pods are not included in residual figures (Table 1). The processing plant wastes, including leaves, pods, and pieces of vines, are handled dry and disposed of on land, but screenings from the wastewater may be fed to animals.

Snap beans (*Phaseolus vulgaris*), usually of the bush type, are machine harvested. In the processing plant the remaining field trash dirt is removed by shaker sieves and updraft blowers. Snipped ends and culled beans are carried in the wastewater and are recovered by screening. Screened solids may be mixed with other wastes and used as feed or field spread.

Table beets (*Beta vulgaris*) are machine harvested and topped in the field. At the processing plant, dirt and remaining leaves are removed. Beets are steam peeled. Fragments, undersize beets, and screened solids form the waste. Only a small portion of beet waste is used as feed.

Cabbage (*Brassica oleracea capitata*) for sauerkraut is harvested both mechanically and by hand. Most of the outer leaves are left in the field; the remaining outer leaves and cores are removed in the processing plant and are usually handled dry. Material spilled during various processes is included in the waste. Only a small amount is used as feed.

Carrots (*Daucus carota*) are mechanically harvested, topped, and often sorted in the field. At the processing plant the carrots are graded and sorted, with small, split, and woody carrots being discarded. The carrots are lye or steam peeled. Trimmed crowns are carried by water and removed by screening. The discarded carrots and screened solids are spread on fields and cattle are allowed to consume the wastes.

Cauliflower (*Brassica oleracea botrytis*) leaf waste has been dehydrated in a pilot-scale alfalfa dehydrator. The dry-matter content of the fresh waste is 7 to 10 percent. The dried meal is separated by air classification into poultry and cattle feed fractions. The poultry meal fraction contains 26 to 31 percent protein and 375 to 620 mg/kg xanthophyll. Individual xanthophyll ratios are similar to those found in alfalfa meal. The poultry meal fraction, fed to chicks at 2 to 10 percent of the diet, was found to be palatable. Pigmentation was comparable to that provided by corn gluten or dehydrated alfalfa meal when fed at similar xanthophyll levels. The cattle feed fraction contained 17 to 20 percent crude protein and 14 to 16 percent fiber (Livingston et al., 1972).

Sweet corn (*Zea mays*) is machine harvested for human consumption, and the stalks and leaves are left in the field. Ears are air cleaned at the processing plant, ends mechanically trimmed, and husks and silks mechanically removed. The unusable portions of the ears are then trimmed away. The corn is washed and kernels are cut from the cob. Broken kernels, silks, and fragments are removed by water-assisted screening, shaker sieves, and air cleaners. These wastes are screened from the wastewater. The cobs, husks, leaves, and stalks are chopped, mixed with screened solids, and ensiled or stacked. Corn waste is sold as feed. The corn waste at one plant contained 90 percent husk and leaf, 8 percent cob, and 2 percent kernel and had 30 to 40 percent dry matter. The corn screening had only 5 percent solids content. The wastewater contains a significant organic load and has potential for recovery by methods similar to those used for potato wastes.

Peas (*Pisum sativum*) are mechanically harvested and vines and pods are removed by mobile or stationary viners and used for feed. At the processing plant the peas are dry cleaned, then washed. Remaining leaves, vines, pods, and pea shells form the waste. During grading, unacceptable peas are removed. The wastewater is screened for solids. The quantity of wastes handled dry or allowed to enter the waste stream varies between plants. Dry and screened wastes are fed to animals or field spread when animal feeding is not feasible.

Pimento (*Pimenta officinalis*) waste consists of pulp, skins, and seeds. The waste is dried and has been found to be rich in red xanthophylls. The xanthophyll content ranges from 896 to 1114 mg/kg. The wastes contain 11.9 percent protein, 9.48 percent fat, and 44.5 percent fiber, on a dry-matter basis. The antioxidant ethoxyquin is effective in preventing carotenoid losses in storage. Levels of 1 percent pimento meal fed in combination with alfalfa meal are considered adequate in poultry diets. Red xanthophylls that contain ketone groups when fed alone impart a pink or reddish color to egg yolks and poultry skin (Livingston et al., 1974).

Pumpkins (*Cucurbita pepo*) are harvested by hand or machine, and vines are left in the field. At the processing plant the pumpkins are washed and the vine ends trimmed. Pumpkins are split or chopped into large pieces and washed by machine; the seeds are removed manually. Unusable pieces are discarded. The pieces are then steamed and pressed; the solids from the pressed liquid may be centrifuged and returned to the product. The pressed pumpkin is pulped and finished. The residual from the finisher contains seeds, skins, and fiber; seeds and fiber are also screened from the wastewater. Some seeds are processed for human consumption, but the other solid wastes are used for animal feed. A significant amount of fine particles passes through screens into the wastewater.

Spinach (*Spinacia oleracea*) and other greens are machine harvested. The dirt is removed by rotating screens, and weeds, discolored leaves, and roots are manually removed and disposed of. These are not used as feed. Following washing, the screened solids are fed to livestock.

ANIMAL BY-PRODUCTS

Dairy Whey

Whey is the largest waste in the dairy industry and results from the production of cheese. Other dairy wastes, which are insignificant by comparison are (1) adulterated milk, which may be disposed of by dumping or irrigating, or may be powdered and used as animal feed; (2) rinsewater from tanks, lines, and equipment containing milk solids, which are recycled into nonfluid food products. Previously rinsewater was disposed of as liquid waste.

Marketing of whey products and economic considerations seem to be the reason whey continues to be a waste problem (Jones, 1974).

Quantity

Whey processing and utilization was reviewed by Clark (1979). Whey production was calculated from data on cheese production in the United States. Manufacture of cheddar cheese results in 9 kg sweet whey/kg cheese produced. Cottage cheese manufacture results in production of 6 kg acid whey/kg cheese produced. In 1976 it was estimated that 15.6 million tons of liquid whey were produced, and about 88 percent of this was sweet whey; the equivalent whey solids were 1,011 million kg. About 57 percent of the whey solids were further processed into human food or animal feed products; 438 million kg were wasted. About 33 percent of the whey solids that were further processed were used in animal feed.

Physical Characteristics

Commercial whey products are numerous and include condensed whey from acid and sweet whey; dry whey products, which include dry whole whey, partially delactosed dry whey, and partially demineralized dry whey; and lactose and whey solids wet blends.

Chemical Composition

Raw whey is dilute and contains 93 to 94 percent water, 0.7 to 0.9 percent crude protein, 0.5 to 0.6 percent fat, 4.5 to 5 percent lactose, 0.2 to 0.6 percent acid, and 0.5 to 0.6 percent ash (Jones, 1974). Dried whole whey contains 13.1 percent crude protein, 76.9 percent lactose, 9.0 percent ash, 0.98 percent calcium, and 0.76 percent phosphorus, dry-matter basis (Schingoethe, 1976).

Nutritional Value

Liquid whey has been used as an animal feed for centuries. It can be fed at levels up to 30 percent of ruminant diets and 20 percent of swine diets, dry-matter basis. Because of its high water content it is usually used near the cheese plants where it is produced. Alternatively, it may be processed into more concentrated forms, but a major problem is the cost of dehydrating, especially at smaller cheese plants. Schingoethe (1976) gave an excellent review of the feeding value of whey and whey products. Condensed whey has been fed to ruminants and swine, and fermented ammoniated condensed whey has been fed successfully to cattle and sheep (Schingoethe, 1976). Dried whey products have been fed to all types of livestock, including poultry; however, poultry have a lower lactose tolerance than other animals, and dried whey should be limited to a maximum of 10 percent of the diet. Levels of 3 to 4 percent dried whey are considered optimal. Ruminants appear to be able to consume very large quantities of liquid or dry whey products.

Results from feeding trials determining the upper limit for feeding whey products to ruminants have varied. Calves can be fed milk replacers containing up to 89 percent dried whey. Dried whey is also useful for improving silage quality. Whey protein concentrates can be recovered from whey; whey proteins are very high quality proteins. The limiting amino acids are phenylalanine and tyrosine. Whey protein concentrates, prepared by various methods, usually contain 50 to 75 percent protein, 20 to 30 percent lactose, and 5 percent or less ash, dry-matter basis. The production of whey protein concentrates leaves a large residual of deproteinized whey. Deproteinized whey has been formed into blocks containing 0.6 percent nitrogen, 70 to 72 percent lactose, and 12 percent ash. The blocks have been used as licks for cattle and calves.

Processing Methods

Bough and Landes (1976) reviewed methods of recovering proteins from whey and investigated the use of chitosan in coagulating whey proteins. Following freeze drying, the coagulated solids contained 72.3 percent protein, 6 percent lactose, 9.5 percent ash, 6.8 percent moisture, and 2.15 percent chitosan. The coagulated solids were fed to rats and the PERs obtained were similar to those for casein and whey solids without chitosan.

Fermented, ammoniated whey (FACW) is formed by fermenting the lactose in sweet or acid whey while maintaining the pH with ammonia to produce ammonium lactate (Juengst, 1979). The FACW is condensed to 55 to 65 percent solids; a 60 percent solids product blends well, handles easily, and is stable. It contains 45 percent crude protein, 37 percent lactic acid as ammonium lactate, and 4.7 percent ash, dry-matter basis. In the past few years research has been conducted on FACW as a feed supplement for dairy and beef cattle. Results of these trials were reviewed by Juengst (1979). In feeding trials, FACW was found superior to urea as a source of nitrogen.

Seafood Processing Wastes

The fish processing industry uses many types of fish and shellfish and is widely spread along coastlines and estuaries. The quantity of waste varies tremendously among plants and among fish and shellfish types, from 0 percent for whole rendered fish to 85 percent for some crabs; the average for all types of seafood wastes is about 30 percent. Soderquist and Williamson (1975) have presented a review of these wastes. Some wastes are recovered in fish meal or fish rendering plants, and there are also numerous by-products that are not used for animal feeds. The remoteness and small size of some plants reduces the feasibility of processing the wastes into by-products.

Quantity

The by-products produced at rendering plants are fish meal, fish oil, and fish solubles. In 1977, 257 million kg fish meal were produced in the United States (Pennington and Husby, 1979). Fish meal is widely used for animal feeds, and its processing and nutritional value has been thoroughly studied. Condensed fish solubles are concentrated from the water removed at fish meal and oil plants. In 1968, 65 million kg fish solubles were produced (Soderquist and Williamson, 1975). Fish solubles are used in feeds and fertilizers.

The technology for processing solid wastes is available but for various reasons is not used, and great quantities of wastes are disposed of on land or at sea. The demand for fish meals is greater than U.S. production can supply.

Nutritional Value

Patton and Chandler (1975) investigated chitinous products by in vivo fermentation. Chitin constitutes 12.3 percent of freshwater crayfish meal, 12.9 percent of crab meal, and 7.6 percent of shrimp meal. Samples of shrimp meal, crab meal, and purified chitin were placed in the rumen of fistulated steers. Average rumen solubilities were 17.4, 35.7, and 21.5 percent, respectively. Solubility of crab meal in the rumen was 15 percent over that in water alone. This research investigating their degradation in the rumen was the first step in determining the feasibility of feeding chitinous materials to ruminants.

Processing Methods

Wastes from processing fish and crabs have been dehydrated and ground to produce fish meal and crab meal. However, for small processors this method of utilization is not feasible. Fish silage has been produced with the addition of formic acid (Backhoff, 1976) or a combination of propionic and formic acids (Gildberg and Raa, 1977). A stable silage was produced by ensiling a 2:1 mixture of fish waste and barley straw with formic and propionic acids (Gildberg and Raa, 1977). Satisfactory ensiling was reported for mixtures of fish waste and corn stover or peanut hulls with the addition of a small amount of molasses (Samuels et al., 1982). Ensiling of crab waste was not as satisfactory.

Poultry Processing Wastes

Poultry by-products are well utilized; therefore, poultry processing and by-product recovery will be briefly summarized. Egg processing, specifically egg breaking plants, have some wastes that are less well utilized.

Eleven percent of the eggs in the United States were processed at egg breaking plants, producing 362.9 million kg liquid egg products in 1972. In 1973 the total egg production in the continental United States was 5,544 million dozen eggs. Egg production and egg breaking plants are widespread throughout the United States, with the highest concentrations occurring in the Southeast and California. Egg breaking plants use surplus or undergraded eggs; maximum production occurs in late spring and early summer.

Shell egg processing plants and egg breaking plants have similar wastes. Egg losses due to shell damage are 9.3 percent from shell egg processing plants and 6 to 10 percent from egg breaking plants. Some waste eggs are collected for use in pet food. Some of the egg content is lost to waste water. Shells and adhering egg content form a large portion of the waste from egg breaking plants. At an egg breaking plant, starting with the in-shell eggs (100 percent), the following products and wastes are derived: edible food product, 78.2 percent; inedible egg product, 3.4 percent; losses to sewer, 6.3 percent; shell, 11 percent; and egg liquid adhering to shell, 1.75 percent.

The quantity of inedible egg product available for sale as feed can be increased by in-plant conservation practices to 85.5 kg per 1000 kg eggs processed. The egg shells from most plants are disposed of in landfills. Egg shells can be dried, but processing them is not economically feasible except at larger plants (Jewell, 1975). Some studies have been done on increasing recovery of by-products from egg breaking plants, especially egg shell wastes and egg contents (Jewell, 1975).

Red Meat Processing Wastes

The red meat industry is well developed in terms of by-product utilization. By-products have been recovered for many years, and efficiency of by-product production is increasing. The term *red meat industry* will be used to differentiate beef, sheep, and hog processing from poultry processing.

The following sections will describe the processes involved in feed by-product recovery and the composition and feed value of underutilized by-products. Subjects covered include: paunch content, wastewater treatment, tannery wastes, and hair.

Paunch Content

Paunch content is the material from the rumen of beef or sheep or the stomach of swine. Beef paunch content is the major waste in the United States and has been studied as a feed source.

Physical Characteristics The paunch content contains undigested feeds, is yellowish brown in color, has an obnoxious odor, and has a water content of 85 percent (Witherow and Lammers, 1976). Because it is poorly utilized, it is a disposal problem to small plants.

Quantity The quantity of paunch content in the United States was estimated as 0.771 billion kg from the 35 million beef cattle slaughtered annually, or by another estimate, 24.5 kg wet weight per animal or 3.8

kg dry weight per animal (Witherow and Lammers, 1976). Some paunch content is dried for feed use, but landfill or field spreading are the most common disposal methods. Methods of disposal have been enumerated by Witherow (1974) and Witherow and Lammers (1976).

Nutritional Value Paunch content varies in composition with feeding practices used prior to slaughter and with the type of processing. In one study, composition was determined on a dry-matter basis as 12.2 percent crude protein, 25 percent crude fiber, 5.2 percent ether extract, 7.9 percent ash, and 49.6 percent nitrogen-free extract (Ricci, 1977). Summerfelt and Yin (1974) claim that the variability of paunch content is less than that of many commonly used feeds and that paunch content from cattle finished on high protein formulated feeds would be even less variable. The range of values for dehydrated paunch content was 12 to 15 percent protein and 12 to 39 percent crude fiber.

Messersmith (1973) studied the feeding value of paunch feed (dehydrated paunch content) with cattle. In one study with high concentrate beef cattle finishing diets, paunch feed had a depressing effect on feed intake when fed as the only roughage. This effect was not seen when the roughage portion of the diet contained both paunch feed and brome hay. In these tests, paunch feed made up 5, 7.5, 10, and 15 percent of the diets. Inclusion of these levels had no significant effect on average daily gain, daily feed consumption, or efficiency of feed conversion. The report indicated that paunch feed had a value similar to that of poor-quality alfalfa hay when fed to ruminants, or 80 percent of the value of grass hay. In the same study, physical and/or chemical treatments were tested on paunch feed, but the response was not as good as the response of other poor-quality roughages.

Summerfelt and Yin (1974) fed paunch feed to catfish at 10, 20, and 30 percent levels. At the 20 percent level or less there was no significant difference in the final weights of pond-reared fish between those fed diets containing paunch feed and those receiving commercial feed. However, fish fed 30 percent paunch feed were smaller. With 10 and 20 percent paunch feed levels, feed costs were similar to the costs for commercial feeds; feed costs per unit of gain were greater with 30 percent paunch feeds.

Processing Methods Paunch content is sufficiently similar to cattle manure that systems developed for processing manure have potential applicability to paunch content. Witherow (1974) described several methods for processing and feeding paunch content, including using paunch content fresh, ensiled, and dried by gas-fired rotary driers, fluid bed driers, solar driers, or by pressing to reduce the water content. Odor during rotary drying can be controlled by installation of a scrubber. One problem that has been encountered is the limited market for the dried feed. Some plants have installed rotary driers and pelletized the feed to improve marketability; they have successfully sold the feed for use with both swine and cattle. Mixing blood with paunch content and drying the mixture yields a product containing 43 percent protein and handles two waste problems at once. Studies have been conducted on the other systems of processing paunch content mentioned above, but it is difficult to determine the extent of implementation of various systems. The two methods of feed use that appear most feasible are drying or ensiling. Ensiling has been tried with paunch content mixed with cornstalks, corn and beet pulp pellets, or combinations of these (Witherow, 1974; Witherow and Lammers, 1976). Some problems resulted because of the limited acceptability of the ensiled materials by cattle, apparently due to the feed's acidity; the addition of sodium bicarbonate seemed to solve the problem.

Meat Processing Wastewater Treatment

There are numerous methods for treating wastewater from meat packing operations; however, only those with feed by-product potential will be described.

Grant (1976) described a process using ion exchange resins to recover proteins from wastewater. He reported that 2 to 5 percent of the total carcass protein is lost in effluents from meat packing plants and poultry plants. The dried protein product recovered contains 68 percent true protein and 3.7 percent ash, dry-matter basis. There are no significant deficiencies in the amino acid pattern. In processing plants where cooking or rendering takes place, fat may become complexed with the protein and will appear in the dried floc protein by-product.

Feeding trials were conducted in New Zealand with dried effluent protein from a meat plant (Grant, 1976). Chicks were fed diets in which 50 percent of the protein was supplied by dried effluent protein. The dried effluent protein was found to be equal in nutritive value to meat meals and casein, slightly inferior to fish meal, and superior to meat and bone meal and protein extracted from grass. Performance of pigs fed 5 percent dehydrated floc proteins was similar to that of control pigs.

Hallmark et al. (1978) reported on the use of lignosulfonic acid (LSA) and dissolved air flotation for recovery of a usable feed protein product. Known as the Alwatech process, this method is being used in several plants and reduces BOD of wastewater by 60 to 90 percent. Solids content of the sludge is 6 to 12 percent. The sludge is neutralized with calcium hydroxide, and surplus blood may also be added. The sludge is dewatered to 30 to 50 percent water and then dried separately or mixed with other meat by-products.

Alwatech protein concentrate usually contains 10 percent LSA, dry-matter basis. Composition varies with the type of effluent treated; meat and fish effluents have been studied. Because the effluent is treated within 1 or 2 hours, the bacteriological condition is excellent. The product has been substituted for 50 percent fish meal or soybean meal in chick and swine diets with satisfactory results. LSA is a purified derivative from wood pulping liquors.

Paulson and Lively (1979) reported on the use of activated sludge treatment and the use of activated sludge as animal feed. The initial solids content of the sludge was 0.3 to 0.6 percent. Pilot studies on thickening sludge by air flotation and centrifugation were conducted. Sludge samples were dried and analyzed. Nitrite and nitrate content were found to be less than 0.1 percent.

The protein content, dry-matter basis, from several samples was 47 to 57 percent. Feeding trials were conducted with rats where sludge replaced up to 100 percent of the dietary protein. Good results were obtained when sludge provided 25 percent of the dietary protein. The rats gained weight when sludge was fed as the only protein source but at a slower rate than normal. This was probably due to the amino acid imbalance of the sludge. The sludge was analyzed for amino acids and found to have a high level of methionine compared to soybean meal.

Tannery Wastes

In recent years the United States has been producing about 40 million cattle hides per year, 50 percent of which are tanned by domestic tanners (M. Komanowsky and J. C. Craig, USDA Eastern Regional Research Center, Philadelphia, Pa., 1979, personal communication). When processed in the tannery only about 55 percent of the weight of the hide goes into making leather. Most of the remaining 45 percent is wasted, 5 to 10 percent as waste hair, 5 to 10 percent as dissolved proteins, 15 percent as fleshings and trimmings, and 15 percent as splits. A fresh cattle hide contains 64 percent water, 33 percent protein, 2 percent fat, 0.5 percent mineral salts, and 0.5 percent other substances. The 33 percent protein is composed of 87.8 percent collagen, 6.1 percent keratin, 5.2 percent nonstructural proteins (albumins, globulins, etc.), and 0.9 percent elastin. To make leather, the tanner removes most of the noncollagenous materials.

Nutritional Value   Wisman and Engel (1961) prepared two tannery by-product meals, referred to as partially hydrolyzed tannery by-product meal and unhydrolyzed, acetone extracted, tannery by-product meal. Both were made from limed hide fleshings. The hydrolyzed meal contained 93.5 percent dry matter, 68.3 percent crude protein, 1.52 percent ether extract, and 23.2 percent ash. The unhydrolyzed meal contained 79.2 percent dry matter, 67.8 percent crude protein, 7.7 percent ether extract, and 4.2 percent ash. The protein of the meals contained 8.1 percent lysine, 20.9 percent glycine, 0.13 percent tryptophan, and 2.4 percent methionine.

The meals were tested in poultry diets as replacements for up to 75 percent of soybean oil meal protein. Diets had constant levels of protein, energy, calcium, and phosphorus. Maximum levels of meals in poultry diets appeared to be 12.5 to 25 percent of the soybean meal protein. Responses to both meals were similar. Adding tryptophan to the diet did not improve growth rates; thus tryptophan was not considered the first limiting amino acid (Wisman and Engel, 1961).

Waldroup et al. (1970) found that chicks fed 2 to 3 percent hydrolyzed leather meal replacing soybean meal performed as well as the controls even without adding supplemental amino acids. Chicks could be fed up to 8 percent leather meal when supplemented with amino acids with no significant differences between treatments. Metabolizable energy content of the leather meal was 2,920 kcal/kg, dry matter basis. Compared to soybean meal, methionine, lysine, and tryptophan were found to be low in leather meal.

Dilworth and Day (1970) conducted a similar experiment and also found that chicks fed 1 to 3 percent leather meal had equal or greater growth than those fed the basal diet, with or without minimum amino acid levels being specified in the diets.

A study was conducted by Knowlton et al. (1976) using hydrolyzed leather scrap to replace 0 to 75 percent of soybean meal crude protein on an isonitrogenous basis in sheep diets. On a dry-matter basis the leather meal contained 75.4 percent crude protein, 1.8 percent ether extract, 18.6 percent ash, and 3.0 percent chromium. Of the many parameters measured in the study, few were affected by the inclusion of hydrolyzed leather meal in the diet at the 50 and 75 percent substitution levels. There was a decrease in apparent digestibility of crude protein at these levels. Hydrolyzed leather meal protein had digestibility values of 81.2, 71.6, and 71.2 percent in the 25, 50, and 75 percent substitution-level diets, respectively. Several reasons suggested for the lower digestibilities were incomplete hydrolysis during processing, heat damage incurred during flash drying, tanning, or a combination of these.

Hair

Hair is available from two sources: from tannery operations that use the "save-hair system" and from slaughterhouse hog hair, which is removed by scalding and scraping.

Nutritional Value Several studies have been conducted on the feeding of raw and processed hog and cattle hair. Moran et al. (1967a,b) found that raw hog hair fed to roosters was very poorly digested, with a metabolizable energy of only 0.58 Mcal/kg on a dry-matter basis. The energy value of the raw hog hair appeared to be derived mainly from the fat content. Hog hair contains 88.1 percent protein, 6.7 percent fat, and 2.2 percent ash, dry-matter basis. Processing under pressure (3.5 kg/cm^2) at 148°C for 30 minutes greatly improved the digestibility. Metabolizable energy was 2.14 Mcal/kg on a dry-matter basis. They found that both hydrogen bonding and disulfide bonding involving cystine were responsible for the low digestibility of protein in raw hair. The cystine content of protein in raw hair was found to be 10 to 15 percent. When the hog hair was processed, cystine was reduced to 3.5 percent of the protein and glycine was noticeably increased from 4.5 to 6.4 percent.

Processing hair is very similar to processing feathers, with a slightly higher temperature being required, 148°C compared to 142°C. In chick-growing diets, up to 5 percent of the soybean protein in a 20 percent protein corn-soybean diet could be replaced by processed hog hair with little effect on growth or feed efficiency. If processed hog hair is substituted for soy protein completely, amino acid supplementation is necessary to prevent severe growth depression. Growth depression was completely overcome by supplementation with lysine, methionine, tryptophan, and glycine, which were the first through fourth limiting amino acids, respectively (Moran et al., 1967a,b).

Moran and Summers (1968) also studied cattle hair in chick diets. The hair had been removed from the hide following calcium hydroxide and sodium sulfide treatment. The metabolizable energy of the raw cattle hair was 1.69 Mcal/kg and was increased to 2.25 Mcal/kg, dry-matter basis, when processed in the same manner as hog hair. The cystine level in raw cattle hair was 5.34 percent of protein and was reduced to 2.92 percent by processing; glycine was increased. Changes in the amino acid content, which had not been observed when processing hog hair or feathers, showed decreases in histidine, lysine, and tyrosine. The authors suggested that this may have been caused by nonenzymatic browning. Substituting the processed cattle hair for soy protein in a 20 percent protein diet did not affect chick performance. However, if processed cattle hair was substituted for all of the soy protein, supplementation with methionine, lysine, tryptophan, histidine, and glycine was required to prevent growth depression.

Kornegay and Thomas (1973) found that diets containing 2 to 3 percent processed hog hair meal could be substituted for soybean meal on a digestible protein basis without depressing growth or feed efficiency. At levels above 6 percent processed hog hair, however, feed intake was depressed. Amino acid deficiencies, imbalance, or poor availability were suggested as causes.

ALTERNATIVE USES FOR FOOD PROCESSING WASTES

Other than for animal feed, uses for food processing wastes have developed because of the increased cost of fossil fuels for energy. Some food processing wastes are now being used for fuel or for alcohol production. Depending upon cost of fuel, some materials (e.g., almond hulls, fruit pits, and nut shells) may be burned as fuel rather than utilized as animal feed. Also, food processing wastes high in fermentable carbohydrates and sugars may be utilized in alcohol production.

ANIMAL AND HUMAN HEALTH PROBLEMS AND REGULATORY ASPECTS

Pesticide Residues

The possibility of harmful pesticide residues must be considered when using crop material wastes. Pesticide use and consequent residues on crops for human consumption are regulated and monitored by federal and state agencies.

Food processors have information on the pesticides used on the crops that they process. It is important that this information be obtained from the processor by those intending to use food processing waste materials for animal feed. In turn, it may often be necessary to analyze the waste material for pesticides to determine that tolerances are not exceeded. Pesticides may be present in higher concentrations on the waste material than on the total raw product received by the food processor. The reason for this is that the pesticide residues are usually on the surface of the commodity and are removed by washing, peeling, and trimming; thereby, they are concentrated in the waste material.

Variable levels of pesticides were reported in apple pomace by Rumsey et al. (1977). Feeding the pomace caused significant accumulation of pesticide in depot fat of pregnant beef cows. Pesticide residues in potatoes were reviewed by McCoy et al. (1975). However, by-product feeds were not covered, only the whole potato and the potato parts. As with most fruits and vegetables, the residue content is higher in the peels. McCoy et al. (1975) state that potatoes, as a root crop, are less likely to carry toxic residues than above-ground crops because the pesticides must either be in the soil or translocated from the aerial part of the plant to reach the tuber.

Feed use of tomato waste may be limited because insecticide levels are often higher than residue standards set for feeds (Schultz et al., 1976, 1977). Toxaphene is one pesticide involved, and the residue can be present in the waxy layer of the skin. Removal of the tomato skin would increase the value of the pomace. The adhering pulp could then be recovered for food use, as has been similarly studied with peel residue from canning.

Heavy Metals

Heavy metals were not found in significant quantities in the biological solids from fruit-cannery activated sludge or the carcasses of animals fed the sludges (Esvett, 1976).

Chromium accumulated in tissues, particularly kidney tissue, and some in fat in chicks fed hydrolized leather meal (Dilworth and Day, 1970). Waldroup et al. (1970) reported that chromium tended to accumulate in kidneys of chicks fed 8 percent leather meal, but not in all tissues examined. Chromium levels are restricted to 2.75 percent in tannery by-products fed to swine (Knowlton et al., 1976).

Animal Health

Apple Pomace and Nonprotein Nitrogen

A serious reproductive problem was encountered when apple pomace was fed with nonprotein nitrogen (NPN) (Fontenot et al., 1977). Apple pomace fed with urea or biuret, or a combination of these, lowered feed consumption and increased body weight losses when compared to corn silage and NPN, and also when compared to apple pomace and protein supplement. Feeding apple pomace and NPN had several serious detrimental effects including high incidence of dead, weak, or deformed calves (Bovard et al., 1977). No explanation has been given for the effects of feeding apple pomace and NPN. Reproductive problems were not encountered when feeding apple pomace with protein supplements. Feeding apple pomace with NPN in any form should be avoided.

Cacao Processing Wastes

The factors that limit cacao processing wastes as feed are theobromine (0.75 to 1.3 percent in shell) and caffeine (Chatt, 1953). Plain chocolate contains 3 percent theobromine and 0.1 percent caffeine (Curtis and Griffiths, 1972). Chatt (1953) reported adverse effects if theobromine intake exceeds 0.025 g/kg body weight. In horses, theobromine intake at the level of 0.027 g/kg body weight has caused death. No waste containing theobromine should be fed to racehorses because it may cause reactions similar to doping.

Cacao shells should only be fed to mature cattle at a maximum level of 2.5 percent of the diet or a maximum of 0.907 kg/day. It should not be fed to pigs, poultry, or calves because the cumulative effect is detrimental (Chatt, 1953). Calves fed 5 to 10 percent chocolate waste exhibited hyperexcitability, exaggerated gaits, and excessively alert appearance; one calf died (Curtis and Griffiths, 1972).

SUMMARY

Many researchers emphasize the variability of food processing wastes. The reasons for this variability are the variability of the raw food material being processed, the differences in production processes employed by different plants, and the different food products produced from the same raw material.

Most of the food processing wastes have substantial nutritional value. A characteristic of most of these wastes is the high moisture, which results in high transportation and dehydration costs per unit of nutrient. Fruit and vegetable processing wastes are generally low in protein and may be limited in energy value; they are probably best suited for feeding to ruminants. Animal processing wastes are generally high in protein, and the protein is usually high in quality. Feeding of wastes usually does not adversely affect animal performance if appropriate levels are included in the diet.

Wastes can be processed by dehydration, but frequently this is not economically feasible. For many wastes, ensiling appears to be feasible. High-moisture materials should be combined with drier materials for good ensiling; the dry materials could consist of poor quality hay or crop residues.

Although caution should be exercised, the feeding of food processing wastes does not appear to pose a serious threat to animal and human health. Pesticide levels need to be monitored. Caution should be used not to feed apple pomace in combination with nonprotein nitrogen to avoid reproductive problems. Wastes such as cacao processing waste, which may contain harmful levels of certain drugs, should be limited to safe levels.

LITERATURE CITED

Amerine, M. A., H. W. Berg, and W. V. Cruess. 1972. Technology of Wine Making. Westport, Conn.: AVI.

Ammerman, C. B., L. R. Arrington, P. E. Loggins, J. T. McCall, and G. K. Davis. 1963. Nutritive value of dried tomato pulp for ruminants. J. Agric. Food Chem. 11:347.

Ammerman, C. B., R. H. Harms, R. A. Dennison, L. R. Arrington, and P. E. Loggins. 1965. Dried Tomato Pulp, Its Preparation and Nutritive Value for Livestock and Poultry. Fla. Agric. Exp. Stn. Bull. No. 691.

Backhoff, H. P. 1976. Some chemical changes in fish silage. J. Food Technol. 11:353.

Ben-Gera, I., and A. Kramer. 1969. The utilization of food industries wastes. Adv. Food Res. 17:77–152.

Bond, J., and P. A. Putnam. 1967. Nutritive value of dehydrated sweet potato trimmings fed to beef steers. J. Agric. 15:726.

Bough, W. A., and D. R. Landes. 1976. Recovery and nutritional evaluation of proteinaceous solids separated from whey by coagulation with chitosan. J. Dairy Sci. 59(11):1874.

Bovard, K. P., T. S. Rumsey, R. R. Oltjen, J. P. Fontenot, and B. M. Priode. 1977. Supplementation of apple pomace with nonprotein nitrogen for gestating beef cows. II. Skeletal abnormalities of calves. J. Anim. Sci. 46:523–531.

Brandt, J. A., B. C. French, and E. V. Jesse. 1978. Economic Performance of the Processing Tomato Industry. Giannini Foundation Information Series No. 78-1. Univ. Calif. Div. Agric. Sci. Bull. 1888.

Brown, A. H., W. D. Ramage, and H. S. Owens. 1950. Progress in processing pear canning waste. Food Packer 31(7):30 and 31(8):50.

Brugman, H. H., and H. C. Dickey. 1955. Potato Pulp as Feed for Livestock. Maine Agric. Exp. Stn. Bull. No. 539.

Brugman, H. H., and H. C. Dickey. 1961. Potato Pulp as a Feed for Livestock. Maine Agric. Exp. Stn. Bull. No. 599.

Burris, M. J., and B. M. Priode. 1957. The Value of Apple Pomace as a Roughage for Wintering Beef Cattle. Va. Agric. Exp. Stn. Res. Rep. 12.

Chatt, E. M. 1953. Cocoa: Cultivation, Processing, Analysis. New York: Interscience.

Clark, W. S. 1979. Our industry today: Whey processing and utilization, major whey product markets—1976. J. Dairy Sci. 62:96–98.

Colston, N. V., and C. Smallwood, Jr. 1972. Waste control in the processing of sweet potatoes. Pp. 85–98 in Proc. Third Natl. Symp. Food Process. Wastes. EPA-R2-72-018.

Cruess, W. V. 1958. Commercial Fruit and Vegetable Products, 4th ed. New York: McGraw-Hill.

Curtis, P. E., and J. E. Griffiths. 1972. Suspected chocolate poisoning of calves. Vet. Rec. 90:313.

Damron, B. L., A. R. Eldred, S. A. Angalet, J. L. Fry, and R. H. Harms. 1974. Evaluation of activated citrus sludge as a poultry feed ingredient. Pp. 142–154 in Proc. Fifth. Natl. Symp. Food Process. Wastes. EPA 660/2-74-058.

Della Monica, E. S., C. H. Huhtanen, and E. O. Strolle. 1975. Stability of protein water concentrates from potato starch factory waste effluents. J. Sci. Food Agric. 26:617–623.

Dickey, H. C., H. A. Leonard, S. D. Musgrave, and P. S. Young. 1971. Nutritive characteristics of dried potato by-product meal for ruminants. J. Dairy Sci. 54:876–879.

Dilworth, B. C., and E. J. Day. 1970. Hydrolyzed leather-meal in chick diets. Poult. Sci. 49(4):1090.

D'Mello, J. P. F., and C. T. Whittemore. 1975. Nutritive value of cooked potato flakes for the young chick. J. Sci. Food Agric. 26:261.

Edmond, J. B., and G. R. Ammerman. 1971. Sweet Potatoes: Production, Processing, Marketing. Westport, Conn.: AVI.

Esselen, W. B., Jr., and C. R. Fellers. 1939. The nutritive value of dried tomato pomace. Poult. Sci. 18(1):45.

Esvett, L. A. 1976. Fruit Cannery Waste Activated Sludge as a Cattle Feed Ingredient. Environ. Prot. Technol. Ser. EPA/2-76-253.

Folger, A. H. 1940. The Digestibility of Ground Prunes, Winery Pomace, Avocado Meal, Asparagus Butts, and Fenugreek Meal. Univ. Calif. Exp. Stn. Bull. 635.

Fontenot, J. P., K. P. Bovard, R. R. Oltjen, T. S. Rumsey, and B. M. Priode. 1977. Supplementation of apple pomace with nonprotein nitrogen for gestating beef cows. I. Feed intake and performance. J. Anim. Sci. 45:513–522.

Gildberg, A., and J. Raa. 1977. Properties of a propionic acid/formic acid preserved silage of cod viscera. J. Food Technol. 28:647.

Gillies, M. T. 1978. Animal Feeds from Waste Materials. Food Technology Review No. 46. Park Ridge, N.J.: Noyes Data Corporation.

Graham, R. P., A. D. Shepherd, A. H. Brown, and W. D. Ramage. 1952. Advanced fruit-waste recovery. Food Eng. 24:82.

Grames, L. M., and R. W. Kueneman. 1969. Primary treatment of potato processing wastes with by-product feed recovery. Water Pollut. Control Fed. 41(7):1358–1367.

Grant, R. A. 1976. Protein recovery from meat, poultry and fish processing plants. In Food from Waste, G. G. Birch, K. J. Parker, and J. T. Worgan, eds. London: Applied Science.

Gray, L. R., and M. R. Hart. 1972. Caustic Dry Peeling of Cling Peaches to Reduce Water Pollution: Its Economic Feasibility. USDA ERS Agric. Econ. Rep. 234.

Guilbert, H. R., and W. C. Weir. 1951. Pear pulp and pear molasses: Nutritional value for cattle and palatability to sheep tested in feeding trials with commercial products. Calif. Agric. 5:6.

Guttormsen, K., and D. A. Carlson. 1970. Status and research needs of potato processing wastes. Pp. 27–38 in Proc. First Natl. Symp. Food Process. Wastes. EPA-12060-04/70.

Hadjipanayiotou, M., and A. Louca. 1979. A note on the value of dried citrus pulp and grape marc as barley replacements in calf fattening diets. Anim. Prod. 23:129–132.

Hallmark, D. E., J. C. Ward, H. C. Isaksen, and W. Adams. 1978. Protein recovery from meat packing effluent. Pp. 288–305 in Proc. Ninth Natl. Symp. Food Process. Wastes. EPA-600/2-78-188.

Hendrickson, R., and W. Kesterson. 1965. By-Products of Florida Citrus, Composition, Technology, and Utilization. Univ. of Florida, Gainesville, Agric. Exp. Stn. Bull. 698.

Hillyer, C. M., and C. T. Whittemore. 1975. Intake by piglets of diets containing cooked potato flake. J. Sci. Food Agric. 26:1215.

Hinks, C. E., D. G. Peers, and I. W. Moffat. 1975. The nutritive value of cooked potato in milk replacers for young calves. J. Sci. Food Agric. 26:1219–1224.

Hinman, N. H., W. N. Garrett, J. R. Dunbar, A. K. Swenerton, and N. E. East. 1978. Tomato pomace scores well as sheep feed. Calif. Agric. 32(8):12.

Hogan, J. M., and M. E. Highlands. 1975. Potato and potato products for livestock feed. Pp. 639–645 in Potato Processing. Westport, Conn.: AVI.

Jewell, W. J. 1975. Egg Breaking and Processing Waste Control and Treatment. Environ. Prot. Technol. Ser. EPA-660/2-75-019.

Jones, H. R. 1973. Waste Disposal Control in the Fruit and Vegetable Industry. Pollution-Technology Review No. 1. Park Ridge, N.J.: Noyes Data Corp.

Jones, H. R. 1974. Pollution Control in the Dairy Industry. Pollution Technology Review No. 7. Park Ridge, N.J.: Noyes Data Corp.

Jones, R. H., J. T. White, and B. L. Damron. 1975. Waste Citrus Activated Sludge as a Poultry Feed Ingredient. EPA-660/2-75-001.

Juengst, F. W., Jr. 1979. Use of total whey constituents—Animal feed. J. Dairy Sci. 62(1):106–111.

Katsuyama, A. M., N. A. Olson, R. L. Quirk, and W. A. Mercer. 1973. Solid Waste Management in the Food Processing Industry. National Canners Association. EPA PB219-019. Available from NTIS.

Knowlton, P. H., W. H. Hoover, C. J. Sniffen, C. S. Thompson, and P. C. Belyea. 1976. Hydrolyzed leather scrap as a protein source for ruminants. J. Anim. Sci. 43(5):1095.

Kornegay, E. T., and H. R. Thomas. 1973. Evaluation of hydrolyzed hog hair meal as a protein source for swine. J. Anim. Sci. 36(2):279.

Livingston, A. L., R. E. Knowles, J. Page, D. D. Kuzmicky, and G. O. Kohler. 1972. Processing of cauliflower leaf waste for poultry and animal feed. J. Agric. Food Chem. 20(1):277–281.

Livingston, A. L., R. E. Knowles, R. H. Edwards, and G. O. Kohler. 1974. Processing of pimento waste to provide a pigment source for poultry feed. J. Sci. Food Agric. 25:483–490.

Livingston, A. L., R. E. Knowles, R. H. Edwards, and G. O. Kohler. 1976. Processing of fresh artichoke trimmings for use in animal feeds. J. Agric. Food Chem. 24(6):1158–1161.

Lofgreen, G. P., and M. Prokop. 1979. Citrus peel liquor as an energy source in cattle growing rations. In Calif. Feeders Day Rep. Univ. Calif. Dept. Anim. Sci., Coop. Ext. Imperial Valley Field Stn.

Mantell, C. L. 1975. Nut Shells and Fruit Pits in Solid Wastes: Origin, Collection, Processing and Disposal. New York: Wiley-Interscience.

McCoy, C. M., and S. E. Smith. 1940. Tomato pomace in the diet. Science 91:388.

McCoy, C. M., J. B. McCoy, and O. Smith. 1975. The nutritive value of potatoes. Pp. 235–273 in Potato Processing. Westport, Conn.: AVI.

Messersmith, T. L. 1973. Evaluation of dried paunch feed as a roughage source in ruminant finishing rations. M.A. thesis. Department of Animal Science, University of Nebraska.

Moran, E. T., Jr., and J. D. Summers. 1968. Keratins as sources of protein for the growing chick. 4. Processing of tannery by-product cattle hair. Poult. Sci. 47(2):570.

Moran, E. T., Jr., J. D. Summers, and S. J. Slinger. 1967a. Keratins as sources of protein for the growing chick. 2. Hog hair, a valuable source of protein with appropriate processing and amino acid balance. Poult. Sci. 46(2):456–465.

Moran, E. T., Jr., H. S. Bayley, and J. D. Summers. 1967b. Keratins as sources of protein for the growing chick. 3. The metabolizable energy and amino acid composition of raw and processed hog hair meal. Poult. Sci. 46(3):548.

Morrison, F. B. 1956. Feeds and Feeding, 22nd ed. New York: Morrison Publishing.

National Research Council. 1976. Nutrient Requirements of Beef Cattle. Washington, D.C.: National Academy of Sciences.

Osten, B. J. 1976. Protein from potato starch mill effluent. P. 196 in Food From Waste, G. G. Birch, K. J. Parker, and J. T. Worgan, eds. London: Applied Science.

Pailthorp, R. E., J. W. Filbert, and G. A. Richter. 1975. Waste disposal. P. 646 in Food Processing. Westport, Conn.: AVI.

Pattee, E. C. 1947. Winery Waste Recovery. Cincinnati: Research Division, National Distillers Products Corp.

Patton, R. S., and P. T. Chandler. 1975. In vivo digestibility evaluation of chitinous materials. J. Dairy Sci. 58:397–403.

Paulson, W. L., and L. D. Lively. 1979. Oxidation Ditch Treatment of Meatpacking Wastes. EPA-600/2-79-030.

Pennington, H., and F. Husby. 1979. University of Alaska fishmeal research. Univ. of Alaska, Alaska Seas and Coasts 7(1):6–8.

Prokop, M. 1979. Dried winery pomace as an energy source in cattle finishing rations. In Calif. Feeders Day Rep. Univ. Calif. Dept. Anim. Sci., Coop. Ext. Imperial Valley Field Stn.

Ricci, R. 1977. A Method of Manure Disposal for a Beef Packing Operation. First Interim Tech. Rep. EPA-600/2-77-103.

Richter, G. A., K. L. Sirrine, C. I. Tollefson. 1973. Conditioning and disposal of solids from potato waste water treatment. J. Food Sci. 38:218–224.

Rogers, C. J., and W. E. Coleman. 1978. Protein from *Aspergillus niger* grown on starchy waste substrate in animal feeds. Waste Mater. Food Technol. Rev. 46:139.

Rosenau, J. R., L. F. Whitney, and R. A. Elizondo. 1976. Low waste water potato starch/protein production process—Concept, status, and outlook. Pp. 118–128 in Proc. Seventh Natl. Symp. Food Process. Wastes. EPA-600/2-76-304.

Rosenau, J. R., L. F. Whitney, and J. R. Haight. 1977. Potato juice processing. Pp. 284–291 in Proc. Eighth Natl. Symp. Food Process. Wastes. EPA-600/2-77-184.

Rosenau, J. R., L. F. Whitney, and J. R. Haight. 1978. Economics of starch and animal feed production from cull potatoes. Pp. 89–99 in Proc. Ninth Natl. Symp. Food Process. Wastes. EPA-600/2-78-188.

Samuels, W. A., J. P. Fontenot, K. E. Webb, Jr., and V. G. Allen. 1982. Ensiling of seafood waste and low quality roughages. VPI and State Univ. Anim. Sci. Res. Rep. 2:175.

Schingoethe, D. J. 1976. Whey utilization in animal feeding: A summary and evaluation. J. Dairy Sci. 59(3):556.

Schultz, W. G., R. P. Graham, and M. R. Hart. 1976. Pulp recovery from tomato peel residues. Pp. 105–117 in Proc. Sixth Natl. Symp. Food Process. Wastes. EPA-600/2-76-224.

Schultz, W. G., H. J. Neumann, J. E. Schade, J. P. Morgan, P. F. Hanni, A. M. Katsuyama, and H. J. Maagdenberg. 1977. Commercial feasibility of recovering tomato peeling residuals. Pp. 119–136 in Proc. Eighth Natl. Symp. Food Process. Wastes. EPA-600/2-77-184.

Skogman, H. 1976. Production of symba-yeast from potato wastes. In Food From Waste, G. G. Birch, K. J. Parker, and J. T. Worgan, eds. London: Applied Science.

Smallwood, C., R. S. Whitaker, and N. V. Colston. 1974. Waste Control and Abatement in the Processing of Sweet Potatoes. Environ. Prot. Technol. Ser. EPA-660/2-73-021.

Smith, T. J. 1976. Dry peeling tomatoes and peaches. Pp. 194–203 in Proc. Sixth Natl. Symp. Food Process. Wastes. EPA-600/2-76-224.

Smock, R. M., and A. M. Neubert. 1950. Apples and Apple Products. New York: Interscience.

Soderquist, M. R., and K. J. Williamson. 1975. Fish and shellfish wastes. In Solid Wastes: Origin, Collection, Processing and Disposal. New York: Wiley-Interscience.

Stabile, R. L., V. A. Turkot, and N. C. Aceto. 1971. Economic analysis of alternative methods for processing potato starch plant effluents. Pp. 185–202 in Proc. Second Natl. Symp. Food Process. Wastes. EPA-12060-03/71.

Stokes, R. D. 1967. An evaluation of current practices in the treatment of winery wastes. M.A. thesis. University of New South Wales, Australia.

Summerfelt, R. C., and S. C. Yin. 1974. Paunch manure as a feed supplement in channel catfish farming. Pp. 246–257 in Proc. Fifth Natl. Symp. Food Process. Wastes. EPA-660/2-74-046.

Treadway, R. H. 1975. Potato starch. P. 546 in Potato Processing. Westport, Conn.: AVI.

Tsatsaronis, G. C., and D. G. Boskou. 1975. Amino acid and mineral salt content of tomato seed and skin waste. J. Sci. Food Agric. 26:421–423.

U.S. Department of Agriculture. 1980. Agricultural Statistics. Washington, D.C.: U.S. Department of Agriculture.

Waldroup, P. W., C. M. Hilliard, W. W. Abbott, and L. W. Luther. 1970. Hydrolyzed leather meal in broiler diets. Poult. Sci. 49(5):1259.

Walter, R. H., J. B. Bourke, R. M. Sherman, R. G. Clark, E. George, Jr., A. B. Karasz, R. Pollman, S. E. Smith, and G. Lake. 1975. Apple pomace in the dairy regimen. Cornell Univ., Geneva, New York. Food Life Sci. 8:12–13.

Whittemore, C. T., A. G. Taylor, I. W. Moffat, and A. Scott. 1975. Nutritive value of raw potato for pigs. J. Sci. Food Agric. 26:255.

Wilson, L. L. 1971. Adipose tissue concentrations of certain pesticides in steers fed apple waste during different parts of the finishing period. J. Anim. Sci. 33:1356–1360.

Wisman, E. L., and R. W. Engel. 1961. Tannery by-product meal as a source of protein for chicks. Poult. Sci. 40(6):1761.

Witherow, J. L. 1974. Paunch handling and processing techniques. Nat. Provis. 10:14.

Witherow, J. L., and S. Lammers. 1976. Paunch and viscera handling. Pp. 37–66 in Workshop (1973) on In-plant Waste Reduction in the Meat Industry, compiled by J. L. Witherow and J. F. Scaief. Environ. Prot. Technol. Ser. EPA-600/2-76-214.

2
Nonfood Industrial Wastes

INTRODUCTION

The intent of this chapter is to examine the type, quantity, and quality of nonfood industrial wastes that are available in North America and to describe the status of available processing technology suitable for converting such industrial wastes to nutritive animal feeds. Typically, nonfood industrial wastes are not used as feedstuffs and as such fall into the category of underutilized resources. While a few nonfood industrial wastes are suitable for direct animal feeding, most are not and require some processing. This processing is needed to (1) achieve nutritional enrichment through synthesis of protein; (2) increase the availability of nutrients through hydrolysis of large-molecular-weight components; (3) change the physical form by concentration, dilution, or entrapment; (4) convert nonutilizable organics to nutritionally useful materials, such as carbohydrates, fats, and fatty acids; or (5) remove or destroy toxic components.

The nonfood industrial wastes to be covered are those derived primarily from the organic chemical and fermentation industries, with some consideration of municipal solid waste. Note will be made of cases in which wastes are currently being used for animal feed, but the focus will be on underutilized materials.

In order for an underutilized waste to be utilizable, it must (1) be nontoxic or capable of being detoxified completely; (2) be available in sufficient quantity at each source to allow for economic recovery; (3) have some nutritive value either before or after processing.

ORGANIC CHEMICAL INDUSTRY

Quantity

The chemical industry in the United States is large and highly diversified. In 1979 the top 50 chemicals produced totalled about 2.56 billion tons and displayed a typical growth rate of about 7.6 percent per year. Of this total, 86 million tons were synthetic organic chemicals, and the remainder were inorganics (Anonymous, 1980). There is relatively little waste utilizable as a feedstuff, either directly or with processing. Most organic waste is burned to provide process heat. When the waste is too dilute to be burned, it is usually too dilute for useful recovery.

The organic chemical industry traditionally has utilized coal, natural gas, and petroleum as primary feedstocks, with heavy reliance on the latter two. However, as a consequence of the rapidly increasing prices of these commodities and the trend to use indigenous energy resources, changes are occurring in the organic chemical industry (Dasher, 1976). These changes are important to consider because they have direct impact on the types and quantities of wastes generated. Increased use of coal and shale oil, which are likely to be processed via gasification or liquefaction, will lead to increased availability of aromatic and fatty acid compounds. Jahnig and Bertrand (1977) described the environmental problems produced by a coal gasification plant with a capacity of 16,000 tons/day that would generate 6,000 tons water/day containing 2.0 to 4.0 g phenolics/liter, 0.5 to 1.5 g fatty acids/liter, and 8.0 to 11.0 g ammonia/liter. This translates to about 18 tons phenolics, 6 tons fatty acids, and 57 tons ammonia/day.

If all of the organic carbon could be converted to microbial single-cell protein having a protein content of 60 percent (Cooney et al., 1980), then about 5,000 tons single-cell protein/year could be produced from a coal gasification plant. Similarly, if coal liquefaction were used, then wastewater generation would be 13,000 tons/day with a composition similar to that from a gasification plant (Magee et al., 1977). While such plants do not exist in the United States today, they are anticipated, and it is clear that major changes in primary energy feedstock will change the availability of a potentially important nonfood industrial waste. Current technology for processing and refining natural gas and petroleum produces little waste.

Another possible change in primary feedstocks in the chemical industry is a shift to using cellulosic biomass, such as crop residue, forest byproducts, or animal waste, as a source of chemicals and fuels. A change from using traditional liquid or gaseous hydrocarbon feedstocks to using solid lignocellulosic feedstocks will cause a major change in waste products. Lignocellulosic biomass is primarily a mixture of cellulose, hemicellulose, and lignin, with some ash, protein, fats, and other minor components. In fermentation processes for converting the biomass to chemicals and fuels, only the cellulose and hemicellulose are consumed, while lignin, along with the other materials, remains as a residue. As the use of biomass develops, there will be increased availability of these residual materials. In addition, there will be large amounts of microbial cell mass associated with the residues. While the cell mass is likely to be used as much as possible as an animal feed, its use will not be without major difficulty because most of the microorganisms that will be produced are not approved for use as feed materials. Analyses of the wastes generated through these changes in technology are, to a large extent, presented in the sections of this report that focus on forest by-products, food processing by-products, and animal waste.

The availability and use of wastes from coal, shale oil, and cellulosic biomass processing will not be dealt with further, since these are not currently underutilized materials, though they may be in the future.

Physical Characteristics

The major organic chemicals derived from primary feedstocks are methane, ethylene, propylene, and aromatics. In 1973 the chemical industry was analyzed by the U.S. Environmental Protection Agency (1973) for the purpose of developing effluent limitations for the industry. The EPA report separates the various chemical products into four processing categories, which are useful for understanding the nature of the wastes from nonfood industries. These categories are (1) continuous nonaqueous processes, (2) continuous vapor-base processes, (3) continuous liquid-phase reaction systems, and (4) batch processes. The processes of interest are primarily those of continuous liquid-phase reaction systems, because these are most likely to involve organic chemical wastes that can be processed to achieve nutritional enrichment. The continuous liquid-phase processes are summarized in Table 3. Batch processes are mostly run on a smaller scale and produce fewer wastes. Nonaqueous as well as vapor-phase wastes are more commonly recovered as an energy source by direct burning. Aqueous wastes are frequently too dilute for burning.

To place the chemical industry and its potential waste in perspective, the list of chemicals and processes in Table 3 is compared with the top 50 organic chemicals (Anonymous, 1980), which represent, in total, about 80 million tons or 93 percent of organic chemicals manufactured. Only those chemicals that are in both category 3, liquid-phase reaction systems, and the top 50 (as noted in Table 3) are considered further. This list provides an indication of those processes that use technology likely to

TABLE 3   Major Chemicals Produced in Liquid-Phase Reaction Systems

Product	Manufacturing Process
Ethanol[a]	Sulfuric acid hydrolysis of ethylene
Isopropanol[a]	Sulfuric acid hydrolysis of propylene
Acetone[a]	Cumene oxidation with cleavage of hydroperoxide in sulfuric acid
Phenol[a]	Raschig process, chlorobenzene process
	Sulfonation process
	Cumene oxidation with cleavage of hydroperoxide in sulfuric acid
Oxo-chemicals	Oxo-process (carbonylation and condensation)
Includes: N-butyl alcohol	
Isobutyl alcohol	
2-ethylhexanol	
Isooctyl alcohols	
Decyl alcohols	
Acetaldehyde	Ethylene oxidation via Wacker process
Acetic acid[a]	Oxidation of LPG (butane)
	Oxidation of acetaldehyde
	Carbonylation of methanol
Methyl ethyl ketone[a]	Sulfuric acid hydrolysis of butene-2, dehydrogenation of sec-butanol
	Oxidation of LPG (butane)—by-product of acetic acid manufacture
Methyl methacrylate	Acetone cyanohydrin process
Ethylene oxide[a]	Chlorohydrin process
Acrylonitrile[a]	Acetylene-HCN process
Ethylene glycol[a]	Sulfuric acid catalyzed hydration of ethylene oxide
Acrylic acid	CO synthesis with acetylene
Ethyl acrylate	Acetylene and ethanol in presence of nickel carbonyl catalyst
	Oxidation of propylene to acrylic acid followed by esterification
	Reaction of ketone with formaldehyde followed by esterification
Styrene monomer[a]	Alkylation of benzene with ethylene, dehydrogenation of ethylbenzene with steam
Adipic acid	Oxidation of cyclohexane/cyclohexanol/cyclohexanone
	Direct oxidation of cyclohexane with air
Terephthalic acid[a]	Oxidation of para-xylene with nitric acid
	Catalytic oxidation of para-xylene
Dimethyl terephthalate[a]	Esterification of TPA with methanol and sulfuric acid
	Vapor phase methylation of phenol
Para-cresol	Oxidation of para-cymene with cleavage in sulfuric acid
Cresylic acid[a]	Caustic extraction from cracked naphtha
Aniline[a]	Nitration of benzene with nitric acid (L.P.), hydrogenation of nitrobenzene

TABLE 3 (continued)

Product	Manufacturing Process
Chloroprene	Dimerization of acetylene to vinyl acetylene followed by hydrochlorination
	Vapor phase chlorination of butadiene followed by isomerization and reaction
Bis-phenol-a[a]	Condensation of phenol and acetone in presence of HCl
Propylene oxide[a]	Addition of propylene and CO_2 to aqueous calcium hypochlorite
	Liquid phase oxidation of isobutane followed by liquid phase epoxidation
Propylene glycol[a]	Hydration of propylene oxide catalyzed by dilute H_2SO_4
Vinyl acetate[a]	Liquid phase ethylene and acetic acid process
Anthraquinone	Catalytic air oxidation of anthracene
Beta naphthol	Naphthalene sulfonation and caustic fusion
Caprolactam[a]	Hydroxyl amine production, cyclohexanone production, cyclohexanone oximation, oxamine rearrangement, purification, and ammonium sulfate recovery
Toluene di-isocyanate	Toluene nitrification, toluene diamine production, HCl electrolysis, phosgene production, TDI production, purification
Silicones	Reaction of silicon metal with methyl chloride
Naphthemic acids	From gas-oil fraction of petroleum by extraction with caustic soda solution and acidification
Ethyl cellulose[a]	From alkali cellulose and ethyl chloride or sulfate
Cellulose acetate[a]	Acetylation of cellulose with acetic acid (followed by saponification with sulfuric acid for diacetate)
Chlorobenzene	Raschig process
Chlorophenol	Direct chlorination of phenol
	From chloroaniline through diazonium salt
Chlorotoluene	Catalytic chlorination of toluene
Hydroquinone	Oxidation of aniline to quinone followed by hydrogenation
Naphthosulfonic acids	Sulfonation of β-naphthol
	Caustic fusion of naphthalene sulfonic acid
Nitrobenzene	Benzene and HNO_3 in presence of sulfuric acid
Amyl acetate	Esterification of amyl alcohol with acetic acid
Amyl alcohol	Pentane chlorination and alkalin hydrolysis
Ethyl ether	Dehydration of ethyl alcohol by sulfuric acid
Ethyl butyrate	Esterification of ethyl alcohol with butyric acid
Ethyl formate	Esterification of ethyl alcohol with formic acid
Tetraethyl lead	Reduction of ethyl chloride with amalgam of Na and Pb
Formic acid	Sodium hydroxide and carbon monoxide
Methyl isobutyl ketone	Dehydration of acetone alcohol to mesityl oxide followed by hydrogenation of double bond
Naphthol	High-temperature sulfonation of naphthalene followed by hydrolysis to β-naphthol

TABLE 3 (*continued*)

Product	Manufacturing Process
Pentachlorophenol	Chlorination by phenol
Sodium pentachlorophenate	Reaction of caustic soda with pentachlorophenol
Toluidines	Reduction of nitrotoluenes with Fe and H_2SO_4
Hydrazine	Indirect oxidation of ammonia with sodium hypochlorite
Oxalic acid	Sodium formate process
Oxalates	Sodium formate process
Sebacic acid	Caustic hydrolysis of ricinoleic acid (castor oil)
Glycerol	Acrolein epoxidation/reduction followed by hydration
	Propylene oxide to allyl alcohol followed by chlorination
Diethylene glycol diethyl ether	Ethylene glycol and ethyl alcohol condensation dehydration
Dichloro-diphenyl-trichloroethane (DDT)	Monochlorobenzene and chloral in presence of sulfuric acid
Pentachloroethylene	Chlorination of acetylene
Methylene chloride[a]	Methane chlorination
	Methanol esterification followed by chlorination
Pentaerythritol[a]	Acetaldehyde and formaldehyde in presence of basic catalyst
Chloral (trichloroacetic aldehyde)	Chlorination of acetaldehyde
Triphenyl phosphate	Phenol and phosphorous oxychloride
Tridecyl alcohol	From propylene tetramer
Tricresyl phosphate	Cresylic acid and phosphorus oxychloride
Amyl alcohol	Chlorination of pentanes and hydrolysis of amyl chlorides
Acrylamide	Acrylonitrile hydrolysis with H_2SO_4
Higher alcohols	Sodium reduction process
Synthetic amino acids	Acrolein and mercaptan followed by treatment with Na_2CO_3 and NaCN
Organic esters	Alcohol and organic acid, H_2SO_4 catalyst
Trialkylacetic acids	Olefins and CO followed by hydrolysis
Fatty acids	Batch or continuous hydrolysis
Lauric acid esters	Esterification of lauric acid
Oleic acid esters	Esterification of oleic acid
Acetophenone	By-product of phenol by cumene peroxidation
Acrolein	Condensation of acetaldehyde with formaldehyde
Ethylacetate[a]	Acetic acid and ethyl alcohol in presence of sulfuric acid
Propyl acetate	Acetic acid and propyl alcohol in presence of sulfuric acid
Acetin (glyceryl monoacetate)	Glycerol and acetic acid
Propionic acid	Carbonylation of ethyl alcohol with CO at high pressure
	Oxidation of propionaldehyde
Fatty alcohol	Reduction of fatty acid with sodium metal
	High pressure catalytic hydrogenation of fatty acids

TABLE 3 (*continued*)

Product	Manufacturing Process
Butyl acetate	Esterification of acetic acid and butyl alcohol in presence of sulfuric acid
sec-butyl alcohol	Hydrolysis of butylene (in H_2SO_4) with steam
N-butyl alcohol	Condensation of acetaldehyde to crotonaldehyde followed by hydrogenation
N-butyl propionate	Esterification of propionic acid with butyl alcohol (H_2SO_4)
Chloroacetic acid	Chlorination of acetic acid
Sodium chloracetate	Esterification of chloroacetic acid
Chloropicrin (nitrotrichloromethane-CCl_3NO_2)	Picric acid and calcium hypochlorite Nitrification of chlorinated hydrocarbons
Thioglycolic acid	Monochloroacetic acid and H_2S followed by neutralization
Adiponitrile	Adipic acid and ammonia
Sodium benzoate	Benzoic acid neutralized with sodium bicarbonate
Sodium sulfoxalate formaldehyde	Zinc hydrosulfite, formaldehyde and caustic soda
Sodium acetate	Neutralization of acetic acid with caustic soda
Tartaric acid	Maleic anhydride and hydrogen peroxide

[a]Denotes those chemicals listed in the top 50 organic chemicals (Anonymous, 1980).

SOURCE: U.S. Environmental Protection Agency (1973).

generate substantial wastes. It is possible to further reduce this list to meaningful terms by examining (1) the raw materials used in manufacture of the major product, (2) the processes, and (3) the products themselves in order to identify waste streams that contain organics suitable for nutritional enrichment by fermentation. A basic premise in this analysis is that none of the wastes from the organic chemical industry will be suitable for direct animal feeding and that fermentation will be required to nutritionally enrich the waste via protein synthesis. At the same time, easily metabolizable compounds will convert the chemicals to a more metabolically usable form and allow conversion of dilute waste streams to solid material (biological cell mass) that is readily recovered.

Nutritive Value

From the brief description of manufacturing processes identified as providing potential underutilized waste (see Table 3), an analysis of process effluents (U.S. Environmental Protection Agency, 1973) and of process flow sheets (Lowenheim and Moran, 1975) was made and is summarized in Table 4.

TABLE 4 Processes Identified as Possibly Having Underutilized NFI Waste

Product	Process	Possible Usable Components in Effluent
Ethanol	Esterification and hydrolysis of ethylene	Ethanol
	Catalytic hydration	
Isopropanol	Esterification and hydrolysis of propylene	Isopropanol, other alcohols
Acetone	Dehydrogenation of isopropanol	Acetone, isopropanol
Phenol	Cumene peroxidation	Phenol, acetone, cumene
Acetic acid	Acetaldehyde oxidation	Acetic acid, formic acid
	Methanol carboxylation	Acetic acid, methanol, propionic acid
	Butane oxidation	Acetic acid, acetone, methanol, formic acid, methylethyl ketone
Terephthalic acid	Oxidation of p-xylene	Acetic acid, xylene
	via propylene chlorohydrin	Propionaldehyde, propylene glycol
Propylene oxide	Oxidation of isobutane	t-butylalcohol, isopentanols
Propylene glycol	Hydration of propylene oxide	Propylene glycol
		Dipropylene glycol
Vinyl acetate	Ethylene and acetic acid	Acetic acid, acetaldehyde
Cellulose acetate	Acetylation	Acetic acid, cellulose
Methylene chloride	Methanol esterification	Methanol
Pentaerythritol	Catalytic	Acetaldehyde, formaldehyde, formic acid, erythritols
Ethylacetate	Catalytic	Ethanol, acetic acid, esters

From the list of organic chemicals, relatively few materials can be called potentially underutilized wastes. These wastes, which generally are in dilute solution (total organic carbon less than 2 percent), are not suitable for direct animal feeding. Furthermore, they require concentration prior to fermentation processing. Some wastes will contain toxic metals and organics that will preclude their use for feeding or make their detoxification difficult.

An alternative to the production of single-cell protein is production of carbohydrates or fat materials for use as a calorie source in animal feeding. Such technology was used during World War II. The process technology is similar for production of single cells for either a protein or calorie source. However, attention here will focus on protein production. It should be kept in mind that processing for calorie production also could be a useful approach.

Processing

Single-cell protein is a generic term for crude or refined sources of protein whose origin is bacteria, yeast, molds, and algae. There have been a number of reviews published on this subject (Davis, 1974; Pirie, 1975), but the most comprehensive treatments of organisms, processes, and nutritional and food technological aspects of utilization are the books based upon two conferences devoted to single-cell protein (Mateles and Tannenbaum, 1968; Tannenbaum and Wang, 1975).

The need for and use of single-cell protein as a protein supplement is well established, and research and development on microbial protein production has been intense for over a decade. There are a number of large plants in operation and several more under construction.

It is interesting that microbial protein produced from nonagricultural raw materials, such as organic chemical wastes, is not dependent on agricultural sources. It is a synthetic yet complete source of food whose composition can be controlled (Cooney et al., 1980) and contains numerous nutrients in addition to protein.

Some forms of single-cell protein have been used as human food for millenia. All fermented foods contain significant quantities of cellular mass as diverse as bacteria, yeast, and fungi. Thus, the use of such organisms as a basic protein food is a logical extension of previous experience.

There is also good scientific evidence that various types of single-cell protein can be useful as additional protein and vitamin sources in animal diets. During the last few years, much data have accumulated about nutritive value and safety of different kinds of yeasts and bacteria grown on chemicals such as n-alkanes and methanol. Technological processes have been developed for industrial production of these products and good estimates for their application are available. These processes consider economic aspects and the safety of single-cell protein use for feeding swine, broilers, and calves. Amino acid patterns, content of nucleic acid and lipids, and data on possible toxic substances have been studied (Tannenbaum and Wang, 1975).

A complete description of individual processes is outside the scope of this chapter. However, it is important to note that single-cell protein production is carried out in intensive processes that permit high-volume production of protein. A typical process flow may appear as shown in Figure 1.

There are several problems in adapting this technology to organic chemical wastes. The first is the need to concentrate the organic waste to levels of 4 to 8 percent usable carbon. The capital investment for a single-cell

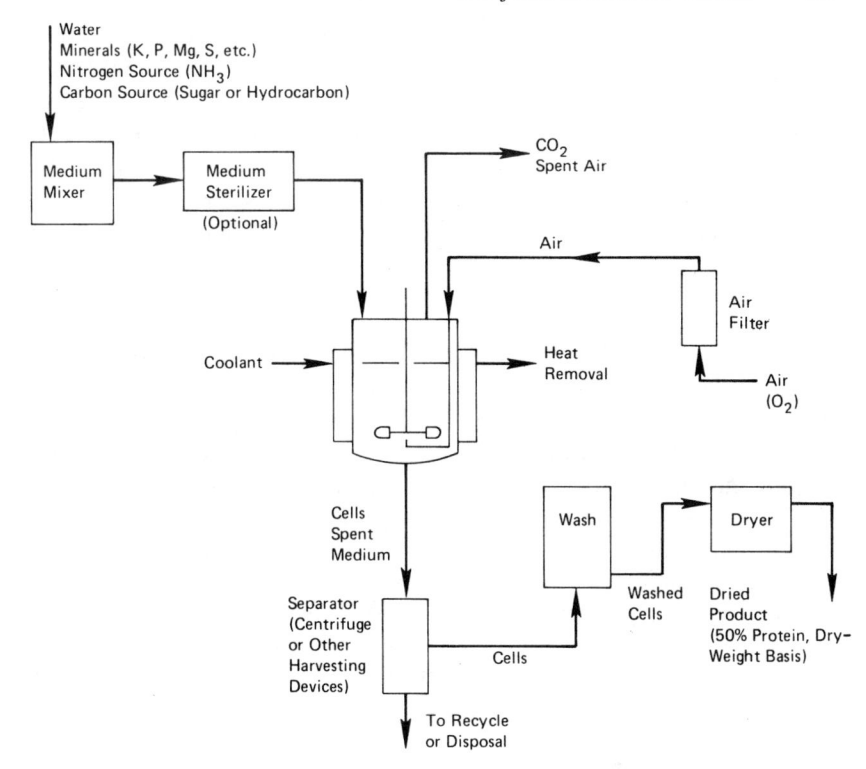

FIGURE 1 Simplified flowsheet of production of single-cell protein. SOURCE: Cooney et al. (1980).

protein manufacturing facility is usually high and it is necessary to maximize both the process productivity and the conversion of organic carbon to cell mass (Cooney, 1975). Second, unlike most single-cell protein fermentations where the substrates are pure, these are variable mixtures of organics. Limited research has been done on the use of mixed substrate fermentation (Silver and Mateles, 1969; Wilcox et al., 1978). Choosing a process, including the organism and the substrate, is extremely complex, and there are many potential processes. In fact, flexibility in process design is an important attribute of single-cell protein. Some important factors, however, will be considered here.

Raw Materials

One of the major advantages of single-cell protein production is the flexibility in being able to choose a variety of organisms able to utilize many different substrates. A brief summary of some substrates considered is presented in Table 5, along with an indication of the typical conversion efficiencies to cell mass. The choice of a carbon source in the design of a process usually depends on factors such as availability, purity, cost, acceptability, and lack of toxicity. In the case of nonfood industrial wastes, the carbon source(s) is fixed by the particular chemical process (see Table 4). It is useful to consider some of the general characteristics of several classes of organic chemicals that serve as substrates for single-cell protein production.

Paraffin Hydrocarbons Paraffin hydrocarbons with 4 to 24 carbons have been of particular interest for microbial protein production. In the production of diesel fuel, it is important to remove the paraffinic fraction to lower the pour point of the oil. Since gas-oil contains approximately 10–40 percent paraffins (the only fraction readily used as a carbon source), it is possible to use microorganisms to dewax the gas-oil. Typically, 1 g of cell mass (dry basis) is produced per gram of paraffin consumed. However, a large unutilized portion of gas-oil passes through the fermentor, creating problems in cell recovery and in removal of residual hydrocarbons from the cells. The removal of residual hydrocarbon is a significant problem in the use of petroleum fractions as carbon sources and requires that the final cell mass be washed extensively with detergent

TABLE 5 Cell Conversion Yields on Various Substrates

Carbon Source	Organism	Temperature (°C)	Cell Yield (g Cell/g Substrate)	Reference
N-paraffins	*Pseudomonas* spp.	30	1.07	Wodzinski and Johnson (1968)
	Nocardia spp.	30	0.98	Wodzinski and Johnson (1968)
	Candida intermedia	30	0.83	Miller and Johnson (1967)
Methane	Mixed bacteria	40	0.62	Sheehan and Johnson (1971)
Methanol	*C. boidinii*	28	0.29	Sahm and Wagner (1972)
	Hansenula polymorpha	37	0.37	Levine and Cooney (1973)
	Mixed bacteria	56	0.42	Snedecor and Cooney (1974)
	Pseudomonas	32	0.54	Goldberg et al. (1976)
Ethanol	*C. utilis*	30	0.68	Johnson (1969)
Glucose	*C. utilis*	30	0.51	Johnson (1969)

solution, and preferably with organic solvents such as hexane and lower-boiling alcohols. Comprehensive reviews of technology applied to hydrocarbons are provided by Davis (1974), Gutcho (1973), and Rockwell (1976).

Alcohols Alcohols, particularly methanol and ethanol, are readily used by bacteria and yeast for microbial protein production (Cooney and Levine, 1972; Kihlberg, 1972). These substrates are water soluble, and leave no residues associated with the cell mass after drying. Since the lower-molecular-weight alcohols have some of the advantages of both hydrocarbon and carbohydrate fermentations, they have significant potential for single-cell protein processes, and are particularly attractive when considered as a food protein source (Cooney, 1975). Higher alcohols are less well used. The cell yield (g cell mass/g alcohol) is typically 0.5 to 0.7 for methanol and ethanol, respectively.

Carbohydrates A variety of carbohydrates can serve as substrates for single-cell protein processes (Tannenbaum and Wang, 1975). Some, such as sulfite liquor and dairy whey, are being used. In addition, it is possible to utilize starch and molasses to produce protein by conventional fermentation technology. The cell yield on carbohydrates is 0.5 g cell mass/g carbohydrate.

Organisms Bacteria, yeasts, and fungi are all being considered for commercial-scale single-cell protein processes; each has advantages and disadvantages. A comparison of protein content of some microorganisms considered for single-cell protein production is provided in Table 6.

The most common organisms used for human food are yeasts (Reed and Peppler, 1973). They are used in foods as vitamin additives and flavoring agents. Their protein concentration is high and they are easier to recover from the fermentation broth than are bacteria. They also have the advantage of availability in the open market and of being produced from a variety of carbohydrate substrates that are themselves recognized as sources of food. Three yeasts, *Saccharomyces cerevisiae*, *Klavaromyces fragilis*, and *Candida utilis* are considered food yeast.

Bacteria have some advanges over other organisms, particularly for animal diets. They have higher levels of protein that is rich in sulfur amino acids. A major technical and economic problem is the high cost of cell recovery because of smaller cell size. However, bacteria are a good long-range prospect because of the potential economic advantage they offer.

Until recently the higher fungi have received relatively little attention in industrial-scale projects (Anderson et al., 1975; Imrie and Vlitos, 1975). The possibility exists for their production in continuous culture on very

TABLE 6 Protein Content of Various Microorganisms

Microorganism	Protein Content (g/g Dry Cell Weight)	Reference
Bacteria		
Pseudomonas methylotropha	0.83	Gow et al. (1975)
Nethylomonas methanolica	0.82	Dostalek and Molin (1975)
Yeast		
Hansenula polymorpha	0.50	Levine and Cooney (1973)
Candida spp.	0.71	Lainé and Chaffaut (1975)
Molds		
Aspergillus niger	0.35	Imrie and Vlitos (1975)
A. oryzae	0.41	Rolz (1974)
Fusarium graminearum	0.66	Anderson et al. (1975)
Algae		
Spirulina	0.64–0.70	Clement (1975)

inexpensive waste carbohydrate sources. The molds might be of considerable interest because of ease of harvesting from fermentation media and their mycelial nature, which provides natural texture.

Utilization Systems

Experimental

A wide variety of experimental systems utilizing microorganisms to convert a chemical to single-cell protein have been examined (Tannenbaum and Wang, 1975). However, these studies have focused more on low-cost methods of protein production than on utilization of chemical wastes. Purified chemicals such as methanol, ethanol, acetic acid, *n*-alkanes, etc. have been used for production of microorganisms to be used in animal feeding studies. Shacklady (1974) reviewed the response of livestock and poultry to yeast and bacterial single-cell protein. Animal feeding trials with microfungi were described by Duthie (1975). Despite the success of these efforts, microbial protein has remained too expensive in comparison to soybean meal and has not been commercialized to a major extent.

Industrial

The industrial production of single-cell protein from chemicals is limited to a few examples. Amoco Foods has a plant to produce 5,000 to 10,000 ton/year of the food yeast *Candida utilis* from ethanol. Imperial Chemical Industries Ltd. has a 70,000 ton/year plant in England for the production of bacteria on methanol. There are also several large plants producing yeast from paraffins in the USSR. In all cases, the product is utilized as a nutritional supplement. As a consequence, the pricing must be such that it can compete with alternative protein commodities.

Animal and Human Health

Pathogens

Unprocessed organic chemical waste is not likely to contain pathogenic microorganisms. However, the processing by fermentation can introduce pathogens. For this reason, guidelines for single-cell protein used in the feeding of animals have been recommended by the International Union of Pure and Applied Chemistry with regard to limits on enterobacteriaceae, salmonella, *Staphylococcus aureus,* clostridia (total), *Clostridium perfringens,* and Lancefield Group D streptococci (Hoogerheide et al., 1979). In addition, guidelines for preclinical and clinical trials for single-cell protein for human consumption have been published by the United Nations Protein Advisory Group (see appendix in Tannenbaum and Wang [1975] for a summary of these guidelines).

Harmful Substances

A major problem in the utilization of organic chemical waste is the presence of toxic organic chemicals and heavy metals; both can concentrate in microorganisms used for conversion of waste to animal feed. In the production of single-cell protein from alkanes containing aromatic compounds, the single-cell protein is solvent-extracted in order to remove any accumulated or residual material prior to animal feeding.

FERMENTATION INDUSTRY

The fermentation industry can be divided into three broad categories: (1) antibiotics and other therapeutic compounds, (2) chemical catalysts (enzymes), and (3) beverages such as wine, beer, and distilled spirits. Wine industry waste is considered in Chapter 2.

Quantity

The antibiotics industry in the United States has annual sales of about $1 billion, and the total amount of antibiotics produced is 10,000 to 20,000 tons/year (U.S. Tariff Commission, 1970). A summary of all the antibiotics produced was presented by Perlman (1978), showing that there are a large number of different products and hence an expected diversity of waste. However, the major fermentation antibiotics, penicillin, cephalosporin, tetracycline, erythromycin, and the aminoglycosides, account for most of the production and produce most of the pharmaceutical fermentation waste.

In the United States there are approximately 25 fermentation plants producing antibiotics and organic acids; most of these are located in the Midwest, the Middle Atlantic region, and Puerto Rico. The primary waste from this industry is the spent mycelia. Unlike chemical industry wastes, pharmaceutical waste is collected in a highly concentrated form, with a high protein content. Estimating a total antibiotics production of 15,000 ton/year and a ratio of 2.5 kg waste/kg antibiotic, it is possible to estimate the yearly production of waste mycelium at 38,000 tons dry material/year. Currently, these wastes are primarily burned, treated in waste treatment systems, or used as fertilizer.

The alcoholic beverage industry (wine, beer, and distilled spirits) is large and produces substantial waste. About 5.5 kg dry brewers grains are recovered for each 117 liters (31 gallon barrel) of beer brewed (Anonymous, 1977a). This means that about 1 million ton/year were available from 5.5 billion liters (176 million barrels) in 1978 for use in animal feeds (U.S. Department of Commerce, 1980). The distilling industry produces about 1 billion liters of distilled spirits at 50 percent ethanol per year (U.S. Department of Commerce, 1980) and in the process generates an estimated 360,000 tons of waste per year. Essentially all of the brewers and distillers wastes are currently utilized for animal feeding and do not represent underutilized materials.

Physical Characteristics

Most of the waste from antibiotic manufacture is the fungal mycelia that is removed from the fermentation broth by filtration. It will typically have a solids content of about 15 percent. Even with 85 percent moisture, the material is a nonflowing cake. Mycelia contains 20–50 percent protein, about 10–30 percent ash (depending on how much fiber aid is used in filtration), and has a C/N ratio of about 10 (on a dry basis).

Nutritive Value

There are very few published data on the nutritive value of antibiotic-producing organisms, primarily because of the low incentive for industry to develop a feed market for its waste mycelia. However, it is possible to compare some of the data available with information on other fungi and attempt to draw some general conclusions. In Appendix 1, Tables 1 through 3, information on several fungi and actinomycetes is presented.

The protein content of the antibiotic producer is generally low (Pfizer, Inc., Groton, Conn., personal communication); it is diluted by a high ash level resulting from use of insoluble inorganic materials for pH control and as a filtration aid.

Doctor and Kerur (1968) conducted rat feeding studies with *Penicillium chrysogenum* in which mycelia was used in combination with peanut meal to provide the total protein in the diet. It was necessary to supplement mycelia with peanut meal in order to make the feed palatable to the rats. These results demonstrated the usefulness of the mycelia in supplementing lysine and threonine, two amino acids which are low in concentration in peanut meal. In addition, the authors refer to unpublished data on the use of mycelia in chick feeding. Earlier studies by Pathak and Seshadri (1965) and Yakinov et al. (1960) also examined *P. chrysogenum* as a protein feed for animals.

There is a substantial amount of knowledge and experience in the use of fermentation wastes from the brewing and distilling industry, and the composition and nutritive value of these by-products is well defined. A recent study of the feeding value of by-products from ethanol production has also been published (National Research Council, 1981) and represents a good source of relevant information.

Processing Technology

One of the limitations in the use of fungi and other mycelial organisms for animal feed is the problem of digestibility. When waste mycelia is processed to allow reuse as a complex nutrient in fermentation, the problem of digestion may be overcome by acid or enzymatic hydrolysis to solubilize the mycelia. This process has the further advantage of permitting the organic material to be concentrated by evaporation to a molasses-like product (Cooney, unpublished results).

The digestion of fungal cell walls has typically been performed using fermentation broth containing extracellular enzymes. It has most often been observed that treated cells are harder to decompose than cell wall suspensions; viable cells are still harder to digest (Kawakami et al., 1972; Matsuo et al., 1967; Okazaki, 1972; Tabata and Terui, 1963).

There are a number of enzymes involved in degrading cell walls since the mycelial cell wall is composed of a mixture of polymers including chitins, β-glucans, lamarins, and peptides; in addition, many of these polymers are cross-linked.

Microbial digestion of cell walls was reported by Sonoda and Ono (1965); they examined the mycelial cake obtained from kanamycin, streptomycin, and penicillin fermentations. They also used the digest from these mycelia to produce methane in an anaerobic digestor.

Utilization Systems

Experimental

Utilization of antibiotic waste has been examined by blending it with other feed components and incorporating it directly into animal diets. However, there are palatability problems (Doctor and Kerur, 1968). The high inorganic content often precludes its direct use in concentrated form. Degradation of the mycelial cell wall to facilitate dehydration has been examined by Ackerman (1975). It was shown that over 50 percent solubilization of the mycelial solids could be accomplished with less than 20 percent loss to carbon dioxide during aerobic treatment. This material was much easier to dehydrate than the whole mycelia.

Industrial

While antibiotic fermentation waste is not used as animal feeds, the dried by-products from the brewing and distilling industries are used widely for animal feeds. Experience with these materials will facilitate the evaluation and utilization of other fermentation residues if they become available.

Health Considerations

One major problem with the use of mycelia from antibiotics manufacturing is the presence of residual antibiotics. In the case of antibiotics used in animal feed (e.g., tetracycline, bacitracin) the whole broth frequently contains cells plus antibiotics that are both dried and fed. In this situation, there is no waste. However, with wastes from processes for antibiotics destined for human use, it is important that residual antibiotics in the mycelia are not permitted to be widely dispersed. Otherwise, there will be selection processes favoring drug resistance (Smith, 1977). It is necessary to remove traces of some antibiotics from the mycelia before it can be fed to animals.

Regulatory Aspects

For the antibiotics industry, a major limitation in the utilization of mycelial wastes is the problem of residual antibiotics. It may be necessary to develop processing methods to eliminate antibiotics if mycelia is to be used for feed. In order to obtain approval from regulatory authorities for use of any feed material, it is necessary to establish standards for the product. This may be very difficult with antibiotic processes because variable raw material quality and variable process operation may cause significant changes in the waste product. Since the primary product of the fermentation has such a high value, there is little economic incentive to alter process operation to assure routine high-quality mycelia. Thus, establishment of such standards may provide a negative incentive to develop the use of antibiotic mycelia as animal feed.

MUNICIPAL SOLID WASTE

Quantity

Municipal Solid Waste (MSW) is often cited as an important and underutilized resource. The total municipal solid waste available through collection in 1975 in the United States was estimated at 68 million dry tons per year (U.S. Department of Energy, 1979). In Canada, it was estimated by Pequegnat (1975) to be 12.7 million dry tons/year. The availability of municipal solid waste is concentrated in large metropolitan areas, and it is anticipated that large quantities will continue to be available. Competitive uses include direct combustion or possible future conversion to alcohol.

Physical Characteristics

The physical characteristics of municipal solid waste are quite variable and depend strongly on the source. There are usually substantial quantities of cellulosic materials, metals, glass, plastics, and dirt. Furthermore, there is no control over the source. The quality of the municipal solid waste fraction used for animal feeding will be determined by the ability of the processor to separate out undesired materials. Belyea et al. (1979) have examined the composition of municipal solid waste that had been fractionated by several methods. Ash content was quite variable and often above the usual level present in farm animal diets.

The bulk density of municipal solid waste is low and, as a consequence, its collection and storage are problems. Not surprisingly, collection costs are a major fraction of the cost of municipal solid waste.

Nutritive Value

Chemical Composition

The composition of shredded and air-classified shredded solid waste is presented in Appendix 1, Tables 1 to 3 (Belyea et al., 1979). The minerals found in major (> 1 percent) amounts are Si, Al, Ca, Mg, and Fe; in minor (0.01-1 percent) amounts are Na, Zn, Pb, Bo, Cr, Ti, and Cu; and in trace amounts are An, Ar, Mn, P, Vn, Mo, Ni, Co, and Bi. An examination of potentially harmful elements suggested that six—Ba, Sn, Se, Pb, Cr, and Cd—may be present in unacceptable levels. Mertens and Van Soest (1971) analyzed a variety of different paper sources; results for the *Washington Post* are shown in Appendix Tables 1 and 2 for comparison. Newspaper is high in indigestible fiber and low in protein and ash.

Nutrient Utilization

A number of studies have been done on the inclusion of municipal solid waste, more specifically newspaper, into animal diets. Kesler et al. (1967) examined the use of waste paper as an absorbant or carrier for molasses in cattle feeding. Compared with controls using corn silage or corn and soybean meal, digestibility of crude fiber was reduced when paper was the absorbant.

The addition of newspapers to diets of growing dairy steers was studied by Daniels et al. (1970). Newspaper was evaluated at 8 and 12 percent levels, replacing 8 percent cottonseed bulk in the control diet. They found no significant differences in rates of gain and feed efficiency. Carcass grade was not affected. It appears that newspaper can be added up to 12 percent of the diet with no detrimental effect. When paper was used at the 20 percent level in lactating dairy cows, however, there was a decrease in milk yield (Mertens et al., 1971).

Processing

An alternative to direct feeding of MSW would be to hydrolyze the cellulase faster by acid or enzymes to produce a sugar syrup. This syrup could be used as a carbohydrate feed or further processed by fermentation to produce products such as single-cell protein. The process of hydrolysis, followed by a solid-liquid separation would be useful in removing undesired materials from the MSW feedstock.

Animal and Human Health

Although municipal solid waste and newspaper have been shown experimentally to be usable as fiber sources in animal diets, there are several serious health concerns. Excess amounts of several minerals were measured in St. Louis municipal solid waste. In addition, Belyea et al. (1979) reported dangerous levels of polychlorinated biphenyls and other toxic compounds. There is no control over the source of waste, and variable levels of harmful materials may occur. Their removal would be expensive. As a consequence, the use of municipal solid waste for animal feed is not considered viable at this time.

RESEARCH NEEDS

The major factors limiting the use of underutilized nonfood industrial waste are the need for nutritional upgrading and the need to remove toxic materials. Therefore, the research needs relating to the problem of nonfood industrial waste utilization should focus on:

1. Evaluation of nonconventional organisms that will utilize chemical process stream wastes and produce microbial proteins. Such an evaluation should include not only effectiveness of conversion to single-cell protein, but also value of the product for use in animal diets.
2. Development of innovative ways to remove or destroy toxic materials from chemical waste streams, municipal solid wastes, and fermentation industry wastes.

SUMMARY

An examination of the nonfood industry to identify underutilized wastes that could be used directly or after further processing for animal feeding has identified a number of areas where better utilization might be achieved. However, there are two major limitations to the use of waste from these nonfood industries: the need and hence expense of processing to achieve nutritional upgrading and the need to remove toxic or otherwise harmful materials from the waste. Problems in nutritional upgrading are associated with factors such as dilute chemical waste stream and variable quantity and quality.

LITERATURE CITED

Ackerman, R. A. 1975. The Degradation of Solid Wastes. M.S. thesis. Massachusetts Institute of Technology.
Anderson, C., J. Longton, C. Maddix, G. W. Scammell, and G. L. Solomons. 1975. The growth of microfungi on carbohydrates. P. 314 in Single-Cell Protein II, S. R. Tannenbaum and D. I. C. Wang, eds. Cambridge, Mass.: MIT Press.
Anonymous. 1977. Animal feeds in the brewing industry. Pp. 5–6 in Proceedings of the U.S. Brewers Association Feed Conference.
Anonymous. 1980. Facts and figures for the chemical industry. Chem. Eng. News. 58(23):33–90.

Belyea, R. L., F. A. Martz, W. McIlroy, and K. E. Keene. 1979. Nutrient composition and contaminants of solid cellulosic waste. J. Animal Sci. 49:1281–1291.

Clement, G. 1975. Producing *Spirulina* with CO_2. P. 467 in Single-Cell Protein, II. S. R. Tannenbaum and D. I. C. Wang, eds. Cambridge, Mass.: MIT Press.

Cooney, C. L. 1975. In Microbial Utilization of C_1-Compounds. Japan: Society for Fermentation Technology. 183 pp.

Cooney, C. L., and D. W. Levine. 1972. Microbial utilization of methanol. Adv. Appl. Microbiol. 15:337–365.

Cooney, C. L., C. K. Rha, and S. R. Tannenbaum. 1980. Single-cell protein: Engineering, economics and utilization in foods. Adv. Food Res. 26:1–52.

Daniels, L. B., J. R. Campbell, F. A. Martz, and H. B. Hedrick. 1970. An evaluation of newspaper as feed for ruminants. J. Anim. Sci. 30:593–596.

Dasher, J. R. 1976. Changing petrochemical feedstocks—Causes and effects. Chem. Eng. Prog. 72:15–26.

Davis, P. 1974. Single Cell Protein. New York: Academic Press.

Doctor, V. M., and L. Kerur. 1968. *Penicillium mycelium* waste as protein supplement in animals. Appl. Microbiol. 16:1723–1726.

Dostalek, M., and N. Molin. 1975. Studies of biomass production of methanol-utilizing bacteria. P. 385 in Single-Cell Protein II, S. R. Tannenbaum and D. I. C. Wang, eds. Cambridge, Mass.: MIT Press.

Duthie, I. F. 1975. Animal feeding trials with a microfungal protein. P. 505 in Single-Cell Protein II, S. R. Tannenbaum and D. I. C. Wang, eds. Cambridge, Mass.: MIT Press.

Goldberg, I., J. S. Rock, A. Ben-Basset, and R. I. Mateles. 1976. Bacterial yields on methanol, methylamine, formaldehyde, and fermate. Biotechnol. Bioeng. 18:1657–1668.

Gow, J. S., J. D. Litchailes, S. R. L. Smith, and R. B. Walter. 1975. SCP production from methanol: Bacteria. P. 370 in Single-Cell Protein II, S. R. Tannenbaum and D. I. C. Wang, eds. Cambridge, Mass.: MIT Press.

Gutcho, S. 1973. Proteins From Hydrocarbons. Park Ridge, N.J.: Noyes Data Corp.

Hoogerheide, J. C., K. Yamada, J. D. Littlehailes, and K. Ohno. 1979. Guidelines for testing of single-cell protein destined as protein source for animal feed—II. Proc. Appl. Chem. 51:2537–2560.

Imrie, F. K. E., and A. J. Vlitos. 1975. Production of fungal protein from carob. P. 223 in Single-Cell Protein II, S. R. Tannenbaum and D. I. C. Wang, eds. Cambridge, Mass.: MIT Press.

Jahnig, C. E., and R. R. Bertrand. 1977. Environmental aspects of coal gasification. Pp. 127–132 in Coal Processing Technology. New York: American Institute of Chemical Engineers.

Johnson, M. J. 1969. Microbial cell yields from various hydrocarbons. Pp. 833–842 in Fermentation Advances, D. Perlman, ed. New York: Academic Press.

Kawakami, N., H. Kawakami, and T. Sato. 1972. Cell wall disintegration with *Physarium* enzyme in *Sahizosaccharomyces pombe*. J. Ferm. Technol. 50:567–711.

Kesler, E. M., P. T. Chandler, and A. E. Branding. 1967. Dry molasses product using waste paper as a base for a possible feed for cattle. J. Dairy Sci. 50:1994–1996.

Kihlberg, R. 1972. The microbe as a source of food. Ann. Rev. Microbiol. 26:427.

Lainé, B. M., and J. Chaffaut. 1975. Gas-oil as a substrate for single-cell protein production. P. 424 in Single-Cell Protein II, S. R. Tannenbaum and D. I. C. Wang, eds. Cambridge, Mass.: MIT Press.

Levine, D. W., and C. L. Cooney. 1973. Isolation and characterization of a thermotolerant methanol utilizing yeast. Appl. Microbiol. 26:982–990.

Lowenheim, F. A., and N. K. Moran. 1975. Industrial Chemicals. New York: John Wiley & Sons.

Magee, E. M., C. E. Jahnig, and C. D. Kalfaldas. 1977. The environmental influence of coal liquefaction. Pp. 16–22 in Coal Processing Technology. New York: American Institute of Chemical Engineers.

Mateles, R. I., and S. R. Tannenbaum, eds. 1968. Single-Cell Protein. Cambridge, Mass.: MIT Press.

Matsuo, M., T. Yasui, and T. Kobayashi. 1967. Cytolytic enzymes from thermophilic microorganisms: I. Cytase production from *Thermoinactinomyces vulgaris*. J. Ferm. Technol. 49:860–868.

Mertens, D. R., J. R. Campbell, F. A. Martz, and E. S. Hildebrand. 1971. Lactational and ruminal response of dairy cows to ten and twenty percent dietary newspaper. J. Dairy Sci. 54:667–672.

Mertens, D. R., and P. J. Van Soest. 1971. Paper composition and estimated nutritive value. Paper presented at National ASAS Meeting.

Miller, T. L., and M. J. Johnson. 1967. Utilization of normal alkanes by yeast. Biotechnol. Bioeng. 8:549–565.

National Research Council. 1981. Feeding Value of Ethanol Production By-Products. Washington, D.C.: National Academy Press.

Okazaki, H. 1972. On cell wall lytic activity produced by thermophilic actinomyces. III. Some properties of the components of the lytic enzyme produced by *Micropolyspora* sp. J. Ferm. Technol. 50:580–591.

Okazaki, H., and H. Iizuka. 1972. On cell wall lytic activity produced by thermophilic actinomyces. I. Lysis of intact cells of various yeasts (including hydrocarbon assimilating yeasts), thermophilic fungi, and green algae. J. Ferm. Technol. 50:405–413.

Pathak, S. G., and R. Seshadri. 1965. Use of *Penicillium chyrogenum* mycelium as animal food. Appl. Microbiol. 18:262–266.

Pequegnat, C. 1975. Economic feasibility of waste as animal feed. P. 26 in Waste Recycling and Canadian Agriculture. Ottawa: Agricultural Economics Council of Canada.

Perlman, D. 1978. The fermentation industries. ASM News. 43(2):82–89.

Pirie, N. W. 1975. Food Protein Sources. Cambridge, Eng.: Cambridge University Press.

Reed, G., and H. J. Peppler. 1973. Yeast Technology. Westport, Conn.: AVI.

Rockwell, P. J. 1976. Single Cell Protein From Cellulose and Hydrocarbons. Park Ridge, N.J.: Noyes Data Corp.

Rolz, C. 1974. Utilization of cane and coffee processing by-products as microbial protein substrates. P. 273 in Single-Cell Protein II. S. R. Tannenbaum and D. I. C. Wang, eds. Cambridge, Mass.: MIT Press.

Sahm, H., and F. Wagner. 1972. Mikrobielle verwettung von methanol. Arch. Mikrobiol. 84:29–42.

Shacklady, C. A. 1974. Response of livestock and poultry to SCP. Pp. 115–128 in Single-Cell Protein, P. Davis, ed. New York: Academic Press.

Sheehan, B. T., and M. J. Johnson. 1971. Production of bacterial cells from methane. Appl. Microbiol. 21:511.

Silver, R. S., and R. I. Mateles. 1969. Control of mixed-substrate utilization in continuous cultures of *Escherichia coli*. J. Bacteriol. 97:535–543.

Smith, H. W. 1977. Antibiotic resistance in bacteria and associated problems in farm animals before and after the 1969 Swann report. Pp. 344–357 in Antibiotics and Antibiotics in Agriculture. W. Woodbine, ed. Boston: Butterworths.

Snedecor, B., and C. L. Cooney. 1974. Thermophilic mixed culture for bacteria utilizing methanol for growth. Appl. Microbiol. 27:1112–1117.

Sonoda, Y., and H. Ono. 1965. Anaerobic digestion of antibiotic waste water. I. On kanamycin, streptomycin, penicillin waste water and mycelium cake. J. Ferm. Technol. 43(11):842–846.

Tabata, S., and G. Terui. 1963. Studies on microbial enzyme active in hydrolizing yeast cell wall. III. On the protease fraction of the yeast cell wall-lytic enzyme of *Streptomyces albido flavus*. J. Ferm. Technol. 43(10):777–782.

Tannenbaum, S. R., and D. I. C. Wang. 1975. Single-Cell Protein II. Cambridge, Mass.: MIT Press.

U.S. Department of Commerce. 1980. Statistical Abstract of the United States, 100th ed. Washington, D.C.: U.S. Government Printing Office.

U.S. Department of Energy. 1979. The Report of the Alcohol Fuels Policy Review—Raw Fuels Availability Reports. Report No. ET-0114/1. Washington, D.C.: U.S. Government Printing Office.

U.S. Environmental Protection Agency. 1973. Development Document for Effluent Limitations Guidelines and Standards of Performance. EPA Report for Contract No. 68-01-1509. Washington, D.C.: U.S. Government Printing Office.

U.S. Tariff Commission. 1970. Synthetic Organic Chemicals. United States Production and Sales. TC Publication 327. Washington, D.C.: U.S. Government Printing Office.

Wilcox, R. P., L. B. Evans, and D. I. C. Wang. 1978. Experimental behavior and mathematical modeling of mixed cultures on mixed substrates. AIChE Symp. Ser. 74:236–240.

Wodzinski, R. S., and M. J. Johnson. 1968. Yields of bacterial cells from hydrocarbons. Appl. Microbiol. 16:1886–1891.

Yakinov, P. A., O. V. Krusser, and Y. E. Koney. 1960. Investigations of *Penicillium chrysogenum* mycelium as a possible protein source for feeding domestic animals. I. Leningr. Khim. Farmatsevt. Inst. 9:212–219.

3
Forest Residues

INTRODUCTION

Wood residues such as wood pulp and bark have been used as energy sources for ruminants during periods of critical feed shortages, but they have never been generally recognized as alternatives for conventional feedstuffs under normal economic conditions (Scott et al., 1969). Although more than 1.5 million tons sulfate and sulfite pulps from spruce, pine, and fir were fed to cattle and horses in the Scandinavian countries during World War II when feed supplies were limited (Hvidsten and Homb, 1951; Nordfeldt, 1951), the feeding of wood pulp to livestock ceased when conventional feedstuffs became available. Feeding of wood-derived feedstuffs in North America has been largely experimental, with the exception of isolated situations in which wood residues have been fed on a commercial scale.

WHOLE-TREE RESIDUE AND FRACTIONS OF WHOLE TREES

Quantity

Historically there has been little need for forest inventory information on quantities of forest biomass beyond information on raw material needed by the forest products industry. As a result, inventories estimated the volume of forest biomass as merchantable boles of commercially important tree species measured from a minimum of 13 cm diameter at breast height

(DBH) to a minimum top diameter of 10 cm. There have been recent attempts to estimate the total annual forest biomass available for use as feed, fuel, and chemical raw material. Some of these estimates assume current commercial forestland area and forest management practices while other estimates assume maximum forest biomass production from fully stocked, intensively managed forestland. A report by the Society of American Foresters (1979) indicates that the annual forest biomass production from present commercial forestland area could double by the year 2035. The estimates of annual forest biomass potential are shown in Table 7. In addition, the report indicates that if 10 percent of the arable land that is currently private forest, pasture, range, and hay land were used for intensive production, up to 240 million dry tons additional annual forest biomass production could be available. If the additional land were managed under short-rotation, intensive-culture biomass farms, the increase could be nearly 450 million dry tons annually (Inman, 1977).

Some studies have been made to measure and to develop methods to estimate the weight of forest biomass for various species, based on DBH

TABLE 7 Estimates of U.S. Aboveground Forest Biomass Potential (millions dry tons)

Source	1970	1976[a]	2002[a]	2010 to 2035 (With Full Stocking)	2035 to 2085 (With Full Stocking and Intensive Management)[b]
Net growth from commercial forest	408	481	544	726	1,088
Mortality[c]	109	109	109	109	109
Other sources[d]	100	100	100	100	100
Harvest for conventional product uses	−177	−181	−236	−236	−290
Total	440	509	517	699	1,007

NOTE: Estimates are derived from data published by Zerbe (1977) and U.S. Department of Agriculture (1977). Moderate industrial demand is projected (Society of American Foresters, 1979).

[a] Net growth estimates from Thomas H. Ellis, USDA Forest Service, Oct. 20, 1978.
[b] Moderate industrial demand is projected to the year 2020, and it is also assumed that intensive management will double growth on 50 percent of commercial forest.
[c] Assumes mortality is recoverable. Intensified management may reduce mortality, but an equivalent volume would be available as live wood.
[d] Land clearing, noncommercial lands, urban tree removal, and urban wastes.

Reprinted courtesy of the Society of American Foresters.

and height measurements or on DBH alone. Whole-tree weight tables for 23 species growing in New York have been published (Monteith, 1979). The results are tabulated by species and DBH, showing fresh and dry weights of the whole tree (aboveground portion), of the entire bole, and of the bole cut off at various top diameters. Monteith also derived prediction equations for green and dry weight of the various components for 10 species and made recommendations for application of these equations to other species. Eleven species of puckerbrush ranging from 1.0 cm to 16.5 cm DBH were sampled and the total fresh and dry weights measured, along with the fresh and dry weights of leaves, branches, and stems (Ribe, 1973). Prediction equations were also developed to calculate the fresh and dry weight, the total, and each component for various DBH of each species. Ek and Dawson (1976) have published the dry-weight yields of *Populus* "tristis #1" grown under short-rotation intensive culture. The results are expressed in terms of dry-weight yields at four spacings of from 0.23 × 0.23 cm to 1.22 × 1.22 m at up to 5 years of growth. After 5 years they reported total aboveground weights at 1.22 × 1.22 m spacings of 49.2 tons/hectare. This total weight yield consisted of 10.76 tons wood, 1.88 tons bark, 24.72 tons wood and bark, 6.32 tons twigs, 5.22 tons leaves, and 0.35 tons other.

Since aspen (*Populus tremuloides* Michx., *P. grandidentata* Michx., and *P. balsamifera* L.) is the species that has most potential as a livestock feedstuff, it should be considered separately. Aspen is the most widespread species in North America. The range is controlled by adequate moisture levels and cool summer temperatures and stretches from Mexico to the Arctic Ocean and from Maine to Alaska. Residue from aspen utilization is usually 100 percent because when aspen is commercially utilized, it is often the only species processed. Commercially important concentrations of aspen exist in the Northeastern, Great Lakes, and Central Rocky Mountain areas of the United States, and the central portions of Canada. The growing stocks in the Rocky Mountain and Lake States areas of the United States are over 100 and 200 million m^3, respectively (U.S. Department of Agriculture, 1976).

Collectibility

Not all of the annual forest biomass potential shown in Table 7 is collectible. Some will be on lands where it is too costly to harvest because of terrain or distance from markets. Some will not be harvested because of environmental, aesthetic, wildlife, and soil fertility considerations. Harvesting as much as possible of the annual growth from forests under the current forest management practices, and especially under intensive forest management practices, requires new and innovative forest harvesting equipment and schemes. Whole-tree harvesting equipment is currently used to obtain pulp-chip-quality wood and a feed-quality by-product. This equipment was developed during the early 1970s to harvest whole-tree chips for the pulp and paper industry. Currently, the main emphasis is to develop equipment to harvest forest biomass for fuel.

Although limited quantities of forest biomass will be available in the future from primary and secondary wood processing plants, biomass availability for livestock feed will most likely be obtained directly from the forest. This material could be a by-product of the system for the harvesting of whole-tree pulp or fuel chips. The by-product would consist mainly of the leaves or needles, twigs, buds, and wood and bark fines. If veneer, pulp, and sawmill logs were harvested, the potential feedstuff would be a by-product of harvesting the tree tops and branches, and it would also contain leaves, needles, twigs, buds, and wood and bark fines. In the case of short-rotation, intensively managed tree farms, the feedstuff could consist of the whole tree, but most likely would be a by-product after separation of stemwood, which would be used for pulp and fuel chips.

The collectibility of aspen depends upon the region of the country. In the Lake States, northeastern areas of the United States, and central Canada, aspen residues from primary wood processing plants are available, as well as whole-tree or portions of whole-tree chips. In the Rocky Mountain areas where less aspen is harvested, only modest amounts of residues from primary processing are available. Whole-tree or portions of whole-tree chips are not available in the Rocky Mountain area because this harvesting method is not yet practiced there.

The primary markets for aspen in the Lake States, northeastern United States, and central Canada are for pulpwood, sawlogs, and composition board. Residue from these uses are bark, which contains up to 50 percent wood, and sawdust that may contain some bark. These residues are collected and, if not currently used, constitute a disposal problem. For a large operation, such as a pulpmill, a sawmill with dry kilns, or a composition-board plant, the bark and other wood residues are usually burned to supply process and space heat. This can require all of the residue wood except in the case of a sawmill with dry kilns. A sawmill usually requires only about one-half of the sawdust and bark to kiln-dry the lumber. The remainder is often used for landfill or sold as cattle bedding, mulch, or fuel. Where whole-tree harvesting is practiced for pulp chips, the segregated fines and bark are put back on the soil. This residue accounts for up to 20 percent of the harvest, and some of it may have value as a feedstuff. If the whole tree is harvested for fuel without further processing, the fines could be separated at the burning site for feedstuff use. If whole-tree aspen is to be processed to fuel pellets (6 mm diameter × 15 mm long), the fines would not be available because they are fuel for the drier; but the fuel-pellet product could be suitable as a feedstuff.

In the Rocky Mountain area, considerable aspen is available. It is not often harvested, however, because it is in scattered stands, usually at elevations of between 2,100 and 3,300 m, and long distances from processing plants and markets.

Physical Characteristics

Many forms of residues are produced from wood processing plants. For instance, sawmill residue consists of sawdust, planer shavings, lumber edge and end trim, slabs cut from the outer portions of the log, and shredded bark. Usually the larger pieces are processed to pulp chips or used for fuelwood or charcoal. The shredded bark, sawdust, and shavings frequently have no markets, but these residues are used increasingly as fuel by the forest products industry.

The particle size of sawdust, planer shavings, and shredded bark depends upon the processing equipment used. Particle size of sawdust is usually less than 6 mm. A sawmill cutting lumber from bark-free aspen logs in the Lake States will produce sawdust with about 10 percent, by weight, of the particles greater than 6 mm, about 40 percent between 6 mm and 10 mm, and 50 percent less than 10 mm. Planer shavings are usually curls of wood 20 to 50 mm long and of various widths. The particle size of shredded bark also depends upon the species and age of the tree. Some barks are stringy, and older barks tend to break into small pieces. Sometimes, especially in the spring, long strips of bark come free of the log. Bark must be processed to achieve a smaller and more uniform particle size for feeding. This can usually be done quite economically with a hammermill.

Forest residues include treetops, branches, and short lengths of logs. These residues can be reduced in size with portable chippers, such as those used to make whole-tree chips. These chips would have to be further reduced in size if they were to be used for livestock feed. The foliage can be separated from wood and bark by air classification at the whole-tree chipping site.

The bulk density of green sawdust is in the range of 250 to 350 kg/m^3. The moisture content of sawdust from a fresh log depends upon species, time of year, and relative amounts of heartwood and sapwood. Usually fresh or green sawdust will contain 40 to 50 percent water. The bulk density of planer shavings can be as low as 100 kg/m^3 for uncompacted material and up to 250 kg/m^3 for compacted shavings. The moisture content of planer shavings usually ranges from 7 to 15 percent because the shavings are usually cut from dried lumber. The bulk density of shredded bark is in the range of 250 to 350 kg/m³ if processed through a hammermill. The moisture content of fresh bark is usually 40 to 50 percent. The bulk density of fresh whole-tree chips, uncompacted, ranges from 300 to 400 kg/m³.

Storage of the residues and whole-tree materials is a problem when the materials are fresh or have a moisture content above 17 percent. Numerous studies have been made on the storage of fresh pulp chips and residue for fuel use because of the decomposition and spontaneous combustion that have occurred in large piles (Springer et al., 1978; White, 1979). Among the variables of storage are the geographic area, species, size of pile, time of year, and particle sizes of materials in the pile. It would be advisable not to store fresh materials intended for use as a livestock feedstuff. If they are stored, it should be in piles less than 6 m high. If they are to be stored in higher piles for more than a few weeks, provisions should be made to monitor the temperature within the pile. The pile should not be covered with plastic, canvas, or heavy snow because they trap heat and cause more rapid internal heating in the pile. A low, uncovered pile will ventilate and maintain low internal temperatures. Storing fresh wood in anaerobic conditions such as a silo would be a suitable method to prevent heating and decay organisms. Storing fresh bark, whole-tree chips, and foliage is more difficult than storing wood in the form of sawdust, chips, and shavings because of the higher amount of nutrients and biologically active materials.

Nutritive Value

Most untreated woods are quite indigestible. Using an in vitro technique, Millett et al. (1970) determined the relative digestibility of 27 species of trees. A summary of the results is shown in Table 8. All of the hardwood species examined showed some degree of digestibility, ranging from a low of 2 percent to a high of 37 percent. Aspen was most digestible, followed by soft maple and black ash. All of the softwoods were essentially indigestible. These results are in general agreement with those Stranks (1959) obtained with a pure culture of *Ruminococcus flavefaciens*. He reported that aspen and ash were relatively digestible in vitro, whereas elm, birch, and basswood required alkali or chlorite pretreatment before they were appreciably digested. Nehring and Schramm (1951a,b) also reported that ash, aspen, maple, and elm were superior roughage sources for sheep, compared to oak and birch.

There is a positive correlation between the digestibility of the bark and the wood of a given species, with the bark usually being more digestible (see Table 8). The rather high content of ether extract in some bark might contribute to its higher digestibility.

As shown in Table 8, maple twigs and buds are more digestible than the stemwood because they contain less lignin and have a higher percentage of digestible protein and ether extract (Nehring and Schütte, 1950). The soft maple buds and twigs had digestion coefficients of about 36 percent, compared to 20 percent for the wood. It is well known that buds and twigs are preferred browse for a number of wild animal species.

TABLE 8 In Vitro Dry-Matter Digestibility of Various Woods and Barks

Substrate	Digestibility[a] Wood (%)	Bark (%)
Hardwoods		
Red alder	2	—
Trembling aspen	33	50
Trembling aspen (groundwood fiber)	37	—
Bigtooth aspen	31	—
Black ash	17	45
American basswood	5	25
Yellow birch	6	16
White birch	8	—
Eastern cottonwood	4	—
American elm	8	27
Sweetgum	2	—
Shagbark hickory	5	—
Soft maple	20	—
Soft maple buds	36	—
Soft maple small twigs	37	—
Sugar maple	7	14
Red oak	3	—
White oak	4	—
Softwoods		
Douglas fir	5	—
Western hemlock	0	—
Western larch	3	7
Lodgepole pine	0	—
Ponderosa pine	4	—
Slash pine	0	—
Redwood	3	—
Sitka spruce	1	—
White spruce	0	—

[a] 96-hour in vitro rumen digestibility. For comparison, the 96-hour digestibility of cotton liners was 90 percent and of a reference alfalfa, 61 percent.

SOURCE: Millett et al. (1970). Courtesy of *Journal of Animal Science*.

Table 9, taken from Millett et al. (1970), shows the water solubility of various hardwood barks and the in vitro digestibility of the various barks. Significant amounts of hydrolyzable oligosaccharides or phenolic glycosides are present in the extracts of aspen and black ash barks. Approximately one-half of the total weight of material in a hardwood-bark water extract of these barks appears to be carbohydrate that could be of nutritional value to the ruminant. It is assumed the remaining portion of the material consists primarily of phenolic compounds.

It is clear that there are large differences in digestibility between tree species. Of the important tree species found in North America, aspen (*Populus* spp.) is by far the most promising as a ruminant feedstuff; thus the emphasis below is on aspen.

Aspen Bark

Chemical Composition The chemical composition of aspen bark can be quite variable. One of the more complete reports on chemical analysis was prepared by Enzmann et al. (1969) (see Appendix Table 1). Crude protein content of aspen bark is usually less than 3 percent on a dry basis. From 5 to 10 percent is ether-extractable. Crude fiber is usually in the range of 40 to 55 percent, and ash can be highly variable, depending upon the amount of soil contamination.

Nutrient Utilization Enzmann et al. (1969) ground and ensiled aspen bark containing 59 percent dry matter. Lambs were fed the ensiled bark

TABLE 9 Water Solubility of Various Hardwood Barks and Extent of Carbohydrate Dissolution

Species	Weight Loss (%)	Free Sugars (%)	Sugars After Hydrolysis (%)[a]	Dry-Matter Digestibility (%)[b]
Aspen	16.0	3.3	6.2	50
Black ash	22.1	6.5	10.0	45
Yellow birch	11.5	5.0	4.8	16
American elm	10.4	5.0	3.6	27
Sugar maple	5.5	1.7	1.9	14

Carbohydrates in Water Extract

[a] 1-hour hydrolysis at 121°C with 4 percent H_2SO_4 followed by neutralization with lime.
[b] 96-hour in vitro rumen digestibility of unextracted bark.

SOURCE: Millett et al. (1970). Courtesy of *Journal of Animal Science.*

in three forms: with no additions; as a mixture of 87 percent bark and 13 percent ground barley; and as a mixture of 87 percent bark, 13 percent ground barley, and 0.025 percent *Aspergillus oryzae* fermentation product on a wet basis. Digestible dry-matter content of the ensiled bark was 31.7 percent, and total digestible nutrient (TDN) content was 36.7 percent on a dry basis. This compares with the TDN content of about 45 to 50 percent for barley, rye, wheat, and oat straw. The mixtures of bark and barley grain had a TDN content of 44 to 48 percent.

Mellenberger et al. (1971) offered goats diets containing 15, 30, 45, or 60 percent of air-dried bark. By regression they determined that the bark had a digestibility of about 50 percent. This experiment used smooth, green bark from freshly cut trees. Bark from older trees may not be as digestible. In a series of unpublished studies from this same laboratory, aspen bark from other sources appeared to have a dry-matter digestibility closer to 30 percent (L. D. Satter, University of Wisconsin, Madison, 1980, personal communcation).

Similar estimates of digestibility of aspen bark have been reported by Gharib et al. (1975). This study was designed to test the effect of bark particle size on digestibility. Lambs were provided diets that contained 20 percent ground corn, 20 percent soybean meal, and 60 percent bark, plus a mineral and vitamin supplement. Dry-matter digestibility was 27.4, 25.7, and 30.3 percent for bark ground through hammermill screens of 0.32, 0.95, and 1.59 cm, respectively.

Animal Performance Relatively little information is available on animal productivity when aspen bark is fed. Enzmann et al. (1969) offered diets containing approximately 37, 53, or 68 percent ensiled aspen bark (dry basis) to 40 kg wethers. The balance of the diet was a mixture of soybean meal and oats. Wethers receiving diets containing 37 and 53 percent aspen bark gained approximately 0.04 kg/day during the 48-day trial. Wethers receiving the 68 percent bark diet lost about 0.04 kg/day. The authors suggested that other feedstuffs, such as poor quality hay or straw, may be better sources of energy during emergency periods than aspen bark.

Fritschel et al. (1976) fed ewes a diet containing approximately 67 to 70 percent aspen bark (dry-matter basis); 8 percent dehydrated alfalfa; 20 percent a mixture of ground shelled corn, oats, and soybean meal; and the balance as mineral, salt, and vitamin mix. The ewes were on the experimental diet for approximately 11 months, and during this time they lambed, lactated, and were bred. The ewes readily consumed the feed and had moderate gains in body weight. Consumption of the diet ranged from about 1.7 kg to 3.0 kg/day, dry-matter basis. During lactation the ewes received an additional 0.5 kg of a grain mix. An average of 1.3 lambs were weaned per ewe. The ewes appeared normal in all respects, and their

TABLE 10 Feedlot Performance of Cattle Fed Pelleted Diets Containing Whole Aspen Tree Material (93 days)

Performance	Alfalfa	12% Aspen	24% Aspen	36% Aspen	48% Aspen	48% Alkali-Treated Aspen
Initial weight (kg)	321	320	318	318	319	319
Final weight (kg)	375	405	425	431	422	422
Average daily gain (kg/day)	0.6	0.9	1.2	1.2	1.1	1.1
Kg feed/kg gain	19.5	13.8	12.1	12.0	11.6	12.0

SOURCE: Singh and Kamstra (1981a).

performance compared favorably with the control group, which was fed hay.

Aspen Sawdust

Chemical Composition Aspen sawdust has a composition similar to that of whole aspen tree. Protein content is less than 2 percent, and acid detergent fiber is usually in excess of 60 percent. Ash content is usually less than 1 percent, unless there is soil contamination, in which case ash content may range up to 10 percent. Klason lignin content is usually between 16 and 22 percent.

Nutrient Utilization The digestibility of aspen sawdust cut from bark-free logs has been determined. Mellenberger et al. (1971) incorporated sawdust into either high-roughage or high-concentrate diets at levels of 0, 20, or 40 percent, air-dry basis. As expected, the overall digestibility of the high-roughage diets was lower than that of the high-concentrate diets. Measures of apparent digestible dry matter decreased linearly with both types of diets as the percentage of aspen sawdust increased from 0 to 40 percent of the diet. Dry-matter digestibility of the sawdust portion of the diet was 41 percent when it was incorporated into a high-roughage diet and 28 percent in a high-concentrate diet.

Whole Aspen Tree

Chemical Composition As with other wood residues, crude protein content of whole-aspen-tree material is very low (see the section, "Forest

Foliage"). Total ash and mineral content is normally low, but can be high as a result of soil contamination (see Appendix Table 1).

Nutrient Utilization Little information is available on the digestibility of whole-aspen-tree material. Digestibility studies of diets containing 0, 20, 40, or 60 percent of ensiled whole-aspen silage suggested that the digestible dry-matter content of aspen silage in an 80 percent grain diet was only 5 percent but approached 37 percent when incorporated into a 40 percent grain diet (Robertson et al., 1971). This latter digestibility figure is about 80 percent of that expected with wheat straw. Singh and Kamstra (1981a) reported dry-matter digestibilities of 51, 52, 54, 60, and 63 percent for total mixed diets containing 0, 12, 24, 36, or 48 percent ground whole aspen tree.

Animal Performance A series of three experiments involving rather large numbers of cattle fed whole-tree material have been reported from South Dakota. In the first of these experiments (Singh and Kamstra, 1981a), 60 Hereford steers, each weighing approximately 320 kg, were allotted to 12 pens of 5 animals each. Mature aspen trees, including bark and leaves, were harvested in summer in the Black Hills region of South Dakota. The trees were chipped and dried to about 10 percent moisture. The dried chips were hammermilled prior to incorporation into complete pelleted diets. The six treatment diets were (1) 93 percent alfalfa (control), (2) 12 percent aspen, (3) 24 percent aspen, (4) 36 percent aspen, (5) 48 percent aspen, and (6) 48 percent aspen treated with 4 percent sodium hydroxide. Soybean meal was incorporated into the pellets at a level equal to two-thirds of the aspen present. A mixture of molasses and mineral and vitamin mix constituted 7 percent of the pellet, with the balance of the pellet supplied by alfalfa. A summary of animal performance in this trial is shown in Table 10. The cattle fed aspen performed well, and actually were more efficient in converting feed to body gain than were the cattle fed alfalfa pellets. It should be pointed out that substantial amounts of soybean meal were present in the aspen treatments, and that much of the animal response may be due to the highly digestible soybean meal. The authors concluded, however, that the aspen had a net energy for maintenance (NE_m) of 1.53 Mcal/kg and net energy gain (NE_g) of 0.48 Mcal/kg. These values are slightly higher than what one would expect from medium-quality hay.

In the second study (Singh and Kamstra, 1981b) 60 Hereford steers, each weighing 327 kg, were divided into six treatment groups and fed to a slaughter weight of 500 kg. The six treatment groups received the following diets: (1) a high-roughage control composed of 93 percent alfalfa; (2) an all-concentrate diet; (3) a high-concentrate diet with 15 percent alfalfa as roughage; (4) a high-concentrate diet with 15 percent aspen as roughage; (5) a diet of 48 percent aspen, 13 percent alfalfa, and 32 percent soybean meal; and (6) a diet of 48 percent aspen, 13 percent alfalfa, 16 percent soybean meal, and 16 percent chicken manure. All diets were fed in meal form.

The results of this trial are shown in Table 11. Animals receiving the high-grain diets (2, 3, and 4) gained the most and had the highest feed efficiency. It would appear from this study that aspen was contributing relatively little to the performance of feedlot cattle fed high-grain diets. Animals tended to eat more of the aspen-diluted diets, but feed efficiency dropped sufficiently with aspen addition to suggest that aspen was contributing little or nothing to the net energy available for growth.

The third experiment (Singh, 1978; Singh et al. 1978) consisted of a field trial involving feeding aspen to wintering beef cows. Three treatment groups of 67 cows each were fed for 25 weeks. The three treatments were (1) mixed grass hay as a control; (2) aspen-alfalfa (60:40) pellets; and (3) 88.5 percent aspen silage, 10 percent corn, 1 percent limestone, and 0.5 percent urea.

The cows were placed in the experiment on November 16. Animals on treatments 1 and 2 were given 0.72 kg protein supplement starting February

TABLE 11 Feedlot Performance of Cattle Fed Complete Mixed Rations Containing Alfalfa and/or Whole Aspen Tree Material (126 days)

Performance	Alfalfa	All Concentrate	Concentrate + 15% Alfalfa	Concentrate + 15% Aspen	48% Aspen + 13% Alfalfa + 32% Soybean Meal	48% Aspen + 13% Alfalfa + 16% Soybean Meal + 16% Chicken Manure
Initial weight (kg)	327	326	327	327	327	326
Final weight (kg)	459	533	529	534	467	432
Average daily gain (kg/day)	1.04	1.65	1.61	1.64	1.11	0.84
Kg feed/kg gain	12.4	5.5	6.0	6.8	9.8	12.4

SOURCE: Singh (1978).

1 and March 26, respectively. To maintain animals after calving, an additional supplement of 2.7 kg corn/head/day was given during May. The silage group needed additional supplementation. After 6 weeks, 2 percent dried molasses was added to the silage diet for 4 weeks. A top dressing of soybean meal at the rate of 1.2 kg/head/day was offered starting December 15 to increase the crude protein content of the silage diet to 7 percent. Soybean meal supplementation was increased to 1.7 kg per head per day from January 27 until termination of the experiment. Corn supplementation of 1.5 kg during the month of February and 2.0 kg thereafter was offered to the silage group. Feeding of the treatment diets and their supplements continued until cows and calves were turned out to open range.

Total feed consumption data were not obtained in this experiment, but dry-matter consumption of the three basal diets was similar. The performance of the cows, divided into young and old groups, is shown in Table 12. All cows lost some weight, with the silage group losing slightly more than the control and pellet group. Normal and healthy calves were born. There were no significant differences in birth or weaning weights among the group fed hay, pellet, or silage rations. Cow and calf losses were 3.5 percent and 0.5 percent, respectively, as compared to 1 percent and 10 percent observed during previous years at the ranch where this field trial was conducted. Although the number of open cows was slightly higher with the control group, this difference was not statistically significant. This series of trials involving large numbers of animals suggests that whole-aspen-tree material is best suited for inclusion in high-roughage, maintenance-type diets. Net energy content of the aspen appeared considerably higher when fed to steers receiving high forage diets than when fed to steers receiving high-grain diets. This observation agrees with the finding of Robertson et al. (1971) that digestibility of whole-aspen silage was very low when incorporated into high-grain diets. The digestibility of most roughages would be depressed when incorporated into high-grain diets.

Forest Foliage

Serious and sustained studies on the use of coniferous foliage in feeds for domestic animals and poultry began in the 1930s. Increased interest in more complete utilization of forest tree components has resulted in considerable forest foliage research, especially in the Soviet Union.

Forest foliage is used primarily as a source of carotene, trace elements, and vitamins in poultry, swine, and cattle diets in the Soviet Union. When used, it typically constitutes 3 to 8 percent of the total diet (Keays, 1976). *Muka* is the Russian term for *flour* or *meal,* and it has become the term

TABLE 12 Performance of Young and Old Cows Fed Wintering Diets Containing Whole Aspen Tree Material

Performance	Control Young[a]	Control Old[b]	Pellet Young[a]	Pellet Old[b]	Silage Young[a]	Silage Old[b]
Initial weight (kg)	439.8	493.0	444.0	490.8	435.6	490.9
Final weight (kg)	411.1	448.9	403.5	449.8	389.5	433.3
Change in weight (kg)	− 28.7	− 44.1	− 40.5	− 41.0	− 46.1	− 57.6
Initial condition score[c]	5.1	5.4	5.2	5.5	5.0	5.4
Final condition score[c]	4.8	5.1	4.3	4.9	4.2	4.5
Number of bred cows	28	23	26	28	24	25
Number of open cows	9	5	4	6	5	4
Birth weight of calves (kg)	35.8	34.3	34.3	34.6	33.6	33.8
Weaning weight of calves (kg)	168.3	163.9	171.7	179.5	164.9	168.7

[a] 2 to 5 years.
[b] 6 to 13 years.
[c] Scored on a scale of 1 to 10 with higher values representing cows in better condition.

SOURCE: Singh (1978).

generally used to describe animal feed derived from tree foliage. Muka can be derived from softwood needles or hardwood leaves. The foliage is usually dried and ground and may be pelleted or fed in loose form. The Soviet Union produced approximately 100,000 metric tons of muka in 1975, and plans were for doubling that by 1980 (Keays, 1976).

Foliage is about 10 percent by weight of the unbarked, full bole for mature softwoods and 25 percent for young softwoods. Corresponding values for hardwoods are 5 and 15 percent. Keays (1976) has estimated that the foliage in the world's forests could produce more than 100 million tons of muka. A conservative estimate of foliage practically available would yield somewhat less, perhaps 10 million tons. Machine harvest of trees results in imperfect separation of foliage from bark and wood. Consequently commercial foliage is usually defined as needles, leaves, twigs, shoots, and branches up to 0.6 cm in diameter. In general, the biomass of commercial foliage available from a tree is approximately double that of the leaves alone (Keays, 1976).

There is growing interest in North America in the short-rotation intensive culture of trees to meet growing demand for pulp and paper products. *Populus* spp. has been given the most emphasis in the north central region because in this location it has repeatedly outproduced other species. *Populus tristis* grown under a short-rotation, intensive-culture system can produce up to 42 metric tons aboveground biomass/ha in 5 years (Isebrands et al., 1979). The commercial equipment presently available for whole-tree processing will convert about 80 percent of the tree into wood chips suitable for pulp manufacture, and 20 percent into leaf, bark, and twig material. This latter fraction has been examined as a ruminant feedstuff. Preliminary observations on the digestibility of this material by goats suggest a digestibility value of about 40 percent (L. D. Satter, University of Wisconsin, Madison, 1979, personal communication).

Chemical Composition   The chemical composition of muka varies with the species of tree used, but the values given in Table 13 may be considered characteristic of commercial muka from *Pinus sylvestris*. Muka contains only about 5 percent protein, and it is quite fibrous. It appears to be a good source of carotene and riboflavin.

Nutrient Utilization   The digestibility of conifer muka by sheep has been estimated by Apgar et al. (1977). Sheep were fed a timothy hay diet containing 25 percent of a mixture of spruce and fir muka. The sheep

TABLE 13   Nutrient Composition of Muka Made From *Pinus sylvestris*

Component	Composition, Air-Dry Basis
	(%)
Dry matter	90.4
Organic matter	84.6
Protein	4.9
Ether extract	4.5
Cellulose	32.1
Nitrogen-free extract	43.0
Ash	5.8
Calcium	0.8
Phosphorus	0.5
	(mg/kg)
Carotene	100–250
Riboflavin	20
Vitamin C	20–2000
Vitamin E	20–40
Manganese	120–350
Zinc	25–35
Copper	10–20
Cobalt	0.5–1

SOURCE: Keays and Barton (1975). Courtesy of estate of John Keays.

readily consumed the muka-timothy hay mixture, but the dry-matter digestibility of the mixture was 58.9 percent compared to 64.1 percent for timothy hay. This suggested that the dry matter digestibility of the muka itself was 43.4 percent or a little less than what would be expected of straw.

Animal Performance Keays and Barton (1975) cite quite a number of Soviet studies in which muka was fed, but it is difficult to evaluate the animal response from the information presented. The general impression is that muka addition to the diet improved animal performance. This conclusion is not supported, however, by studies conducted in North America with growing chicks. Gerry and Young (1977) replaced ground corn with 5 percent fir (*Abies* spp.) muka, spruce (*Picea* spp.) muka, or a combination of the two in a starter and finisher broiler diet. At 7.5 weeks of age the broilers given the muka diets had significantly lower body weights. Feed consumption was not different from the control, but the feed:gain ratio was significantly poorer when spruce muka was added alone or in combination. Hunt and Barton (1978), using broiler chicks up to 4 weeks of age, also found that spruce muka incorporated at levels of 2.5 to 10 percent in basal growing diets resulted in lower growth rates.

The North American studies with muka in broiler diets have not indicated any benefit. The high-fiber content of muka does lower the energy value of diets for poultry. In addition, muka made from some tree species, such as spruce, contains phenolic compounds and tannins, both of which may contribute to depression in feed intake and growth (Hunt and Barton, 1978).

Processing Methods

Braconnot (1819) first discovered that cellulose could be converted to fermentable sugars by means of concentrated mineral acid. Research on wood hydrolysis did not progress until Simonsen (1898) suggested using dilute sulfuric acid. Ten to fifteen years later, three commercial-scale wood hydrolysis plants were constructed in the United States. Two plants used dilute sulfuric acid hydrolysis, with the major product being ethyl alcohol, and the other plant used anhydrous sulfur dioxide gas and produced feed (Kressman, 1922).

Research to utilize wood in animal feeds began at the Forest Products Laboratory in 1920 when eastern white pine (*Pinus strobus*) and Douglas fir (*Pseudotsuga menziesii*) sawdust was hydrolyzed and fed to animals at the University of Wisconsin and the U.S. Department of Agriculture,

Beltsville, Md. The work was started as a result of high feed-grain prices during 1918 and 1919. Wood was hydrolyzed and the washings and hydrolyzate were neutralized, concentrated, mixed with the unhydrolyzed residue, and dried (Sherrard and Blanco, 1921).

This type of material was used in several feeding experiments with sheep and dairy cows (Archibald, 1926; Morrison et al., 1922; Woodward et al., 1924). Results indicated that certain animals could eat diets containing up to one-third hydrolyzed sawdust mixture. Animals such as dairy cows requiring considerable energy intake could eat up to 15 percent of the hydrolyzed mixture without noticeable effects on milk production. It was determined that the eastern white pine mixture was 46 percent digestible and that the Douglas fir mixture was 33 percent digestible. It was concluded that feeding hydrolyzed wood was practical only when natural feed grains were in short supply.

Research on wood hydrolysis was conducted in the 1940s to produce concentrated sugar solutions suitable for stock and poultry feed. Over 200 tons of molasses were produced in pilot plants and sent to universities, agricultural experiment stations, and other agencies for feeding tests with milk cows, beef cattle, calves, lambs, pigs, and poultry (Lloyd and Harris, 1955; U.S. Department of Agriculture, 1960). In general, the tests indicated that wood-sugar molasses is a highly digestible carbohydrate feed and was comparable to blackstrap molasses. In addition, torula yeast was grown on neutralized dilute wood hydrolyzate, and the yeast was evaluated as an animal feed ingredient (Harris et al., 1951).

Results from feeding tests with wood molasses led to commercial production during the early 1960s of a concentrated hemicellulose extract called Masonex, a by-product from hardboard production (Turner, 1964).

Various chemical, biochemical, and physical treatments have been suggested to increase the digestibility of wood by animals. These treatments include:

- Hydrolysis with various acids to solubilize the cellulose
- Alkali and ammonia treatment to saponify the ester bonds and promote swelling beyond the waterswollen dimensions to increase microbiological penetration into cell wall structure
- Breaking of lignin-cellulose chemical bonds or actual delignification with various chemicals to yield a digestible cellulose
- Comminution to very small particle size to alter the crystalline structure of cellulose
- High energy electron irradiation to break chemical bonds of the lignin and cellulose

Hydrolysis

Early research and commercial methods increased cellulose digestibility by hydrolysis. This resulted in the production of a molasses product or a product containing a mixture of the hydrolyzed cellulose and the residue after hydrolysis. Operation of commercial production plants was short-lived, however, because of high costs. At the present time in the United States, three plants produce a wood molasses (hemicellulose extract) from wood as a by-product of the production of hardboard by the wet process. Two of these plants use a 1- to 2-minute high-pressure steam cook after which the steam pressure is suddenly released (Turner, 1964). The other plant uses a steam-pressurized refiner to defibrate the wood chips. In each plant the solubilized wood materials, which are mainly hemicellulose sugars and organic acids, are neutralized and either concentrated or spray dried for use in animal diets. The combined production of this molasses and dried product is about 68,000 tons/year on a dry-matter basis. Since this is a by-product, production depends on the markets for hardboard. Typical analysis of this molasses is 0.5 percent protein, fat, and fiber; 6 percent ash; 0.8 percent phosphorus; 3 percent calcium; 34 percent total sugar; and 35 percent moisture.

In addition to the 3 plants presently producing a molasses product, 13 additional plants produce hardboard or medium-density fiberboard by a wet process and do not concentrate solubilized wood sugars and acids. These plants could supply an estimated 180,000 tons, dry-matter basis, of additional molasses. One pulpmill, using a prehydrolysis prior to pulping, is also capable of producing an estimated 27,000 tons molasses/year, dry-matter basis.

Steaming is also an approach to increasing the digestibility of wood. While steaming generally involves no added acid, the cleavage of acetyl groups early in the process provides an acidic medium conducive to hydrolytic action. Steaming has been applied to both straw and wood. Recent applications to wood are described in studies by Bender and colleagues (Bender et al., 1970; Heaney and Bender, 1970). These showed that steam treatment of aspen (*Populus* spp.) chips for about 2 hours at 160° to 170°C gave a product that is readily accepted by sheep at up to 60 percent of the diet and provided normal weight gains and carcass yields. This research led to the process developed by Stake Technology Ltd., Toronto, to make a steamed product from wood and agricultural residues (Bender, 1979; Stake Technology, Ltd., 1979). Steaming time is a few minutes, after which the pressure is rapidly reduced. Two plants in the United States are using this process to produce feed from aspen (Timber Resources, Inc., Bangor, Maine, and Enfor Feeds, Aitkin, Minnesota).

Another process has been developed by Iotech Corp., Ltd., Kanata, Ontario, to steam and explosively release the pressure to produce a wood product that has increased digestibility (Marchessault and St. Pierre, 1978). A hydrolysis process was developed by Jelks (1976) to produce animal feedstuffs. This process is based on additions of a mineral acid in the presence of a catalyst and oxygen. Several plants have been constructed in the United States to produce animal feedstuffs from wood and agricultural residue.

The Madison process using dilute sulfuric acid could also be used if additional molasses is needed (Lloyd and Harris, 1955; U.S. Department of Agriculture, 1960). Over 40 plants based on a modified Madison process are operating in the Soviet Union to make wood sugars for fermentation to yeast for human food or animal fodder. Yeast is a desirable end product from the wood sugars because of the protein value and ability of the yeast to metabolize both 5- and 6-carbon sugars. Some of the wood sugars are fermented to ethanol.

Alkali and Ammonia

Although alkali swelling of wood has not been used commercially to prepare feed, the relative effects of alkali on wheat straw and on poplar (*Populus* spp.) wood were studied by Wilson and Pigden (1964). Both materials were soaked in increasing amounts of sodium hydroxide up to 15 percent of the material weight, and the digestibilities of the products were obtained for straw and wood (see Figure 2); however, the digestion coefficient of the straw was always about 30 percent greater than that of the wood, the maximum digestibilities being about 80 percent for straw and 50 percent for wood. Thus, while the overall reaction mechanism of the alkali appears to be the same, the advantage of straw resides in the greater initial availability of the carbohydrates. A further demonstration of the effect of alkali on the digestion of wood was provided by Pew and Weyna (1962). By alternately swelling 80-mesh spruce (*Picea* spp.) wood in cold 2 N sodium hydroxide and following this by digestion with *Trichoderma viride* enzyme, 80 percent of the total carbohydrates were converted to sugars. Without the alkali treatment, the maximum attainable digestion was only a few percent.

Data are presented relative to the in vitro digestibility and lignin content of some typical hardwood species treated with 1 percent sodium hydroxide at a 20:1 solution-to-wood ratio (see Table 14). All wood samples were ground through a 40-mesh screen in a Wiley mill. Treatment effectiveness was monitored by a Tilley and Terry in vitro rumen technique (Mellenberger et al., 1970) augmented by animal feeding trials with the more promising products.

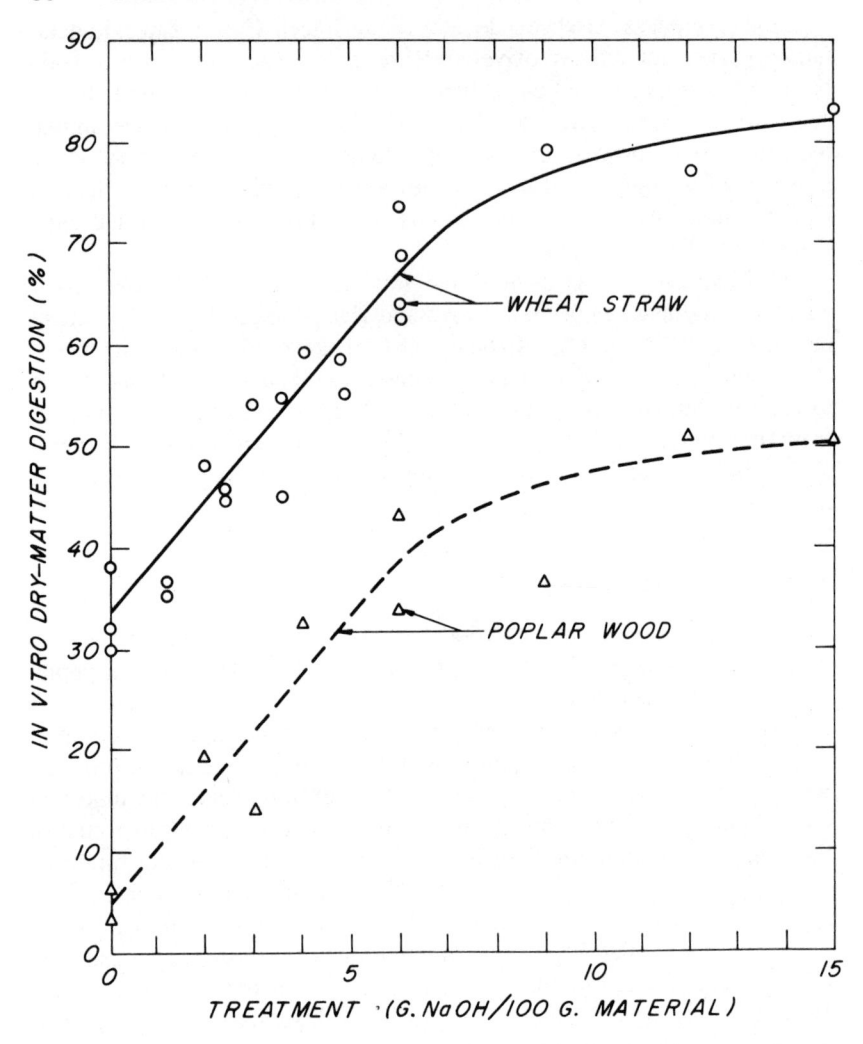

FIGURE 2 Effect of NaOH pretreatment on the in vitro digestion of straw and poplar wood. SOURCE: Wilson and Pigden (1964).

No uniformity exists in the response of the various species to sodium hydroxide treatment. For example, while the digestibility of basswood increased from 5 to 56 percent, the digestibility of American elm (*Ulmus americana*) increased from 9 to only 14 percent. Softwoods, as a group, exhibited digestibilities in the 1 to 5 percent range and were essentially unresponsive to alkali treatment. This difference in response appears to be related to the lignin content of the wood (see Figure 3). A similar

TABLE 14   Effect of NaOH Treatment on the In Vitro Digestibility of Hardwoods[a]

Species	Lignin Content (%)	Digestibility Untreated (%)	Treated (%)
Quaking aspen (*Populus tremuloides*)	20	33	55
Bigtooth aspen (*Populus grandidentata*)	20	31	49
Black ash (*Fraxinus nigra*)	20	17	36
American basswood (*Tilia americana*)	20	5	55
Paper birch (*Betula papyrifera*)	21	8	38
Yellow birch (*Betula alleghaniensis*)	21	6	19
American elm (*Ulmus americana*)	23	8	14
Silver maple (*Acer saccharinum*)	18	20	41
Sugar maple (*Acer saccharum*)	23	6	28
Red oak (*Quercus rubra*)	24	3	14
White oak (*Quercus alba*)	23	4	20

[a] 5 g wood treated 1 hour with 100 ml 1 percent NaOH, washed and dried.

SOURCE: Baker et al. (1975).

relationship appears to apply to other lignocellulosic materials, as evidenced by the common observation that, as forages mature, their lignin content increases and the digestibility of their cellulose and hemicellulose components by ruminants decreases.

To better define conditions for optimum processing, Feist et al. (1970) and Millett et al. (1970) investigated the influence of alkali concentration on the extent of in vitro digestion of aspen and red oak (*Quercus rubra*). The results (see Figure 4) show that from 5 to 6 g NaOH/100 g wood are necessary for maximum effect. This level of alkali is about equivalent to that required to react with acetyl and carboxyl contents of the woods. This led Tarkow and Feist (1969) to postulate that the main consequence of alkali treatment is the saponification of intermolecular ester bonds, thus promoting the swelling of wood beyond waterswollen dimensions and favoring increased enzymatic and microbiological penetration into the cell wall fine structure. Wood substance lost by water extraction at this level of treatment is about 5 percent for aspen and can be ascribed largely to the removal of saponified acetyl groups.

Verification of the in vitro results was obtained in a feeding trial with goats (Mellenberger et al., 1971). Hammermilled aspen sawdust was treated with 0.5 percent NaOH at a 10:1 liquid-to-solid ratio for 2 hours at room temperature. After draining, washing once with water, and air drying, the

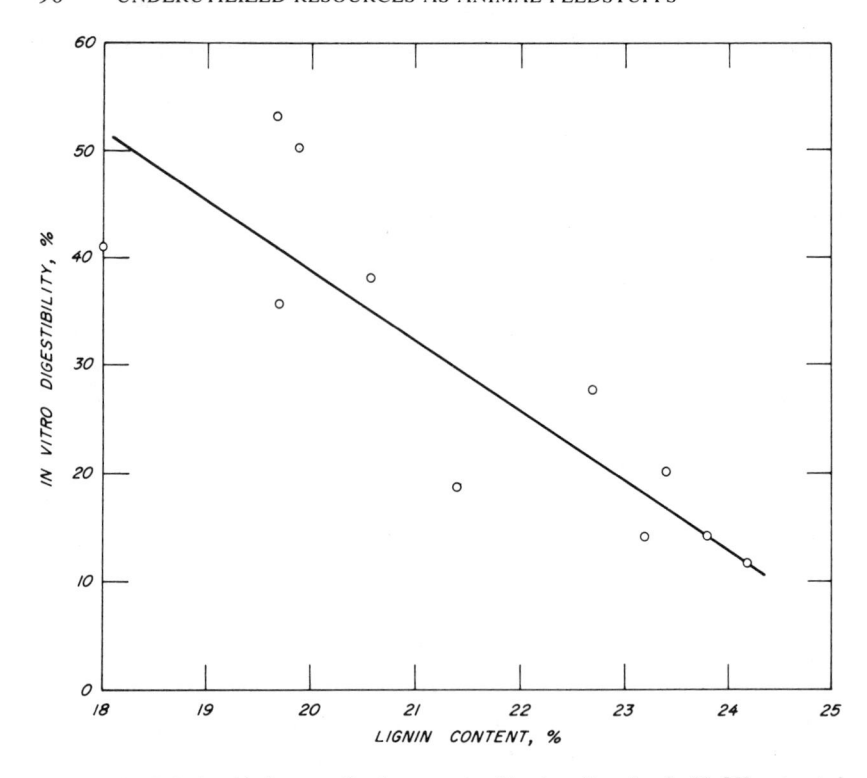

FIGURE 3 Relationship between lignin content and in vitro digestion for NaOH pretreated hardwoods. SOURCE: Feist et al. (1970).

product was incorporated into pelleted diets at levels of 0 to 60 percent. A similar set of diets was prepared from untreated sawdust. Overall diet digestibility, calculated from feed and feces analysis, is plotted as a function of aspen content (see Figure 5). Extrapolation of the curves to 100 percent aspen yielded dry-matter digestibilities of about 41 percent for untreated aspen and 52 percent for the alkali-treated aspen. Alkali treatments can thus increase ruminant utilization of aspen wood by approximately 25 percent, making it equivalent to a medium-quality hay as a source of dietary energy.

Another long-standing approach to upgrading the feeding value of lignocellulosic materials involves treatment with aqueous or gaseous ammonia. It has long been recognized that liquid ammonia exerts a strong swelling action on wood and cellulose and can effect a phase change in the crystal structure from cellulose I and cellulose III (Segel et al., 1954; Tarkow and Feist, 1969). In an attempt to exploit these effects, a number of woods were exposed to anhydrous ammonia in both liquid and gaseous

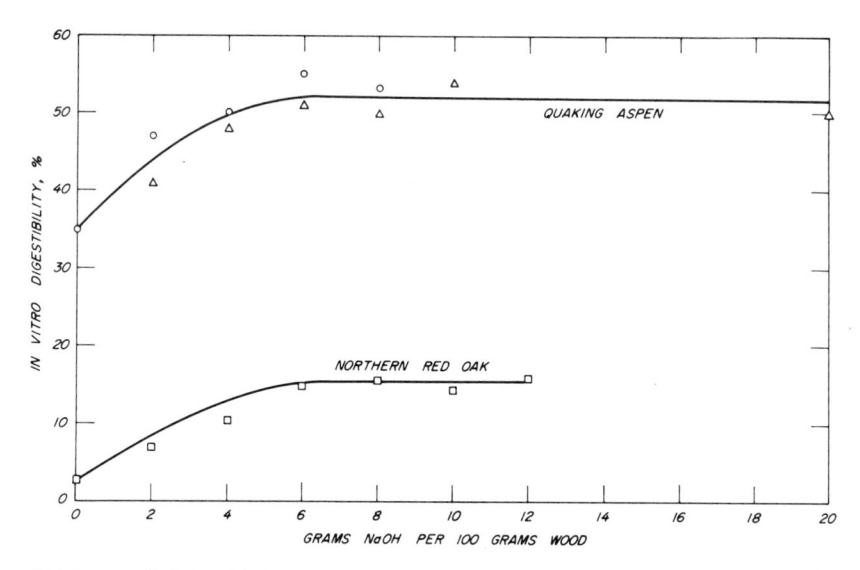

FIGURE 4 Relationship between level of NaOH pretreatment and in vitro digestion for quaking aspen and northern red oak (aspen: ○ = 2 hours in 0.5 percent NaOH; Δ = 1 hour in 1 percent NaOH; red oak: □ = 4 hours in 1 percent NaOH). SOURCE: Feist et al. (1970).

form and the results assayed by changes of in vitro digestibility (Millett et al., 1970). Aspen appeared to be unique in the extent of its response to ammoniation, attaining a digestibility coefficient of about 50 percent as contrasted with Sitka spruce (*Picea sitchensis* (Bong.) (Carr.)) and red oak (*Quercus rubra*), whose ammoniated-product digestibilities were only 2 and 10 percent, respectively.

In an extension of this work, 363 kg ammonia-treated aspen were prepared by subjecting air-dried aspen sawdust to a 2-hour pressure treatment with anhydrous NH_3 gas at 482 kPa and 25°C. After blowdown and removal of excess NH_3, the product was blended into pelleted diets at levels of 0 to 50 percent and fed to goats, with ready acceptance over a 6-week feeding period. From extrapolation of the curve for ration digestibility in relation to wood content, an estimated digestibility of approximately 46 percent was obtained for the ammonia-treated product. This is about 6 percent lower than the in vivo digestibility of the NaOH-treated aspen but still sufficient to rank NH_3-treated aspen in the same range as hay as a potential source of energy. As an added benefit, the NH_3 treatment would provide some nitrogen. As with sodium hydroxide, Tarkow and Feist (1969) described the pertinent action of ammonia treatment to the hydrolysis, ammonolysis in this case, of glucuronic acid-ester crosslinks,

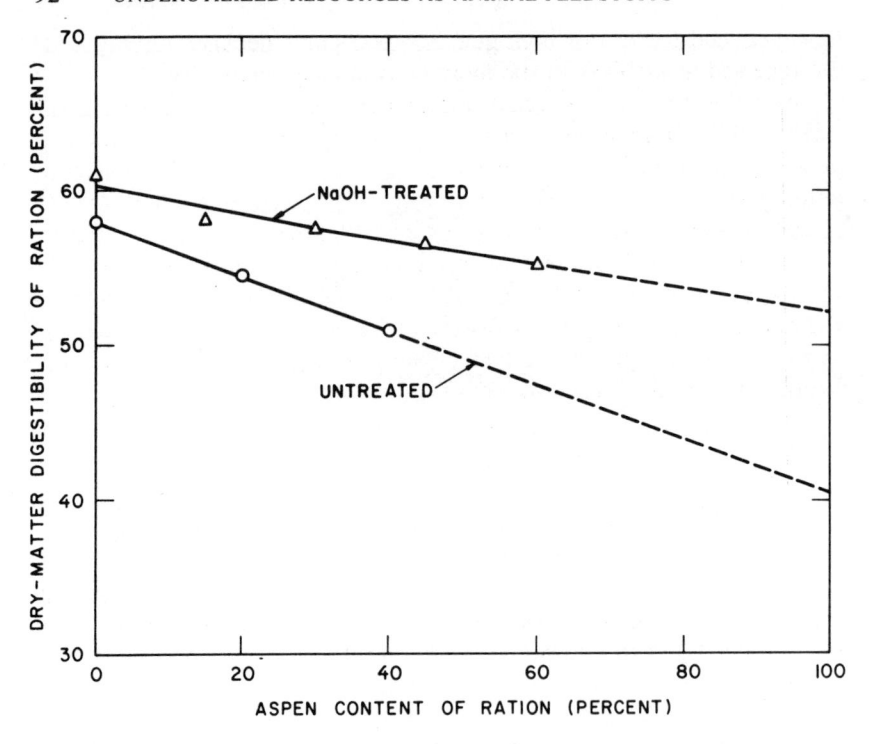

FIGURE 5 In vitro dry-matter digestion of rations containing untreated and NaOH-treated aspen. SOURCE: Mellenberger et al. (1971).

thereby providing more ready access to structural carbohydrates by rumen bacteria and their associated enzymes. Similarly, the wide response-range exhibited by the various wood species toward NH_3 treatment can be ascribed, in part at least, to the extent of lignification.

Delignification

Because lignin is the major roadblock to widespread utilization of the carbohydrate content of the abundant lignocellulosic residues, delignification would appear to provide a straightforward solution to increasing the digestibility of lignocellulosics. That it can be is indicated by the more than 1.5 million tons of sulfate and sulfite wood pulps from pine, spruce, and fir consumed by cows and horses during World War II in the Scandinavian countries (Edin, 1940; Edin et al., 1941; Hvidsten, 1940; Nordfeldt, 1951; Saarinen et al., 1958). This was an emergency situation, however, and the use of high-quality wood pulps declined rapidly with the renewed availability of the more conventional feedstuffs.

Since the presence of lignin appeared to result in low digestibility, Baker (1973) determined what degree of delignification was required to obtain carbohydrate utilization by the rumen microflora. By selecting pulping variables, a series of kraft pulps was prepared having a broad range of residual lignin contents whose digestibilities were then determined in vitro. Four wood species were included in the investigation: two hardwoods, paper birch (*Betula papyrifera*) and red oak; and two softwoods, red pine (*Pinus resinosa*) and Douglas fir.

As shown in Figure 6, an appreciable difference exists in the delignification-digestibility response between hardwoods and softwoods. With

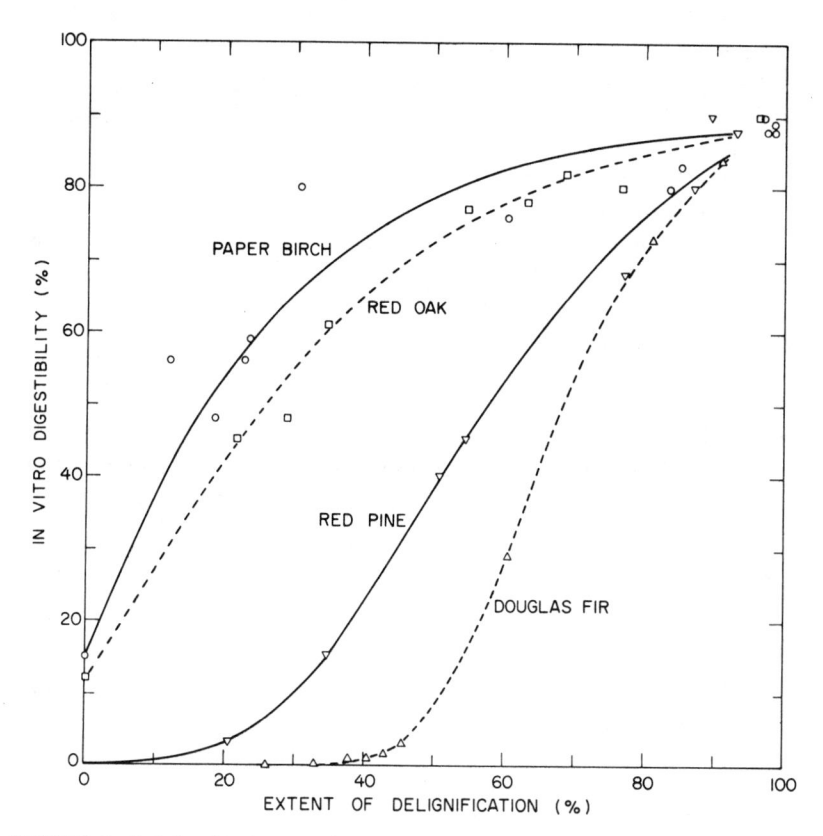

FIGURE 6 Relationship between in vitro digestibility and extent of delignification for kraft pulps made from four wood species. SOURCE: Baker (1973).

the two hardwoods, digestibility increases rapidly with delignification and approaches a plateau of about 90 percent as delignification approaches completion. With the two softwoods there is a distinct lag phase, especially pronounced with Douglas fir, during which extensive delignification is accompanied by only minor increases in digestibility. Following this lag phase, digestibility rises rapidly and almost linearly with delignification up to the digestibility maximum.

As interpolated from these four curves, the extent of delignification necessary to obtain a product having an in vitro digestibility of 60 percent, that of a good-quality hay, is shown in Table 15 along with data on the lignin content of the original woods. As with alkali treatment (see Figure 3), digestibility response strongly correlated with lignin content, the response being measured in terms of the degree of lignin removal needed to achieve a specified level of product utilization. Additional support for this lignin dependence was obtained by Saarinen et al. (1959) in an investigation of the in vivo digestibility of a series of birch and spruce pulps prepared by 10 different pulping techniques. Recalculation of their data to fit the format of Baker (1973) provided the results shown in Figure 7, which also includes Baker's curves for red pine and paper birch for comparison. In spite of the wide variation in delignification techniques employed by the two investigations, their results are quite comparable. This leads to the further conclusion that it is primarily the degree of delignification that governs pulp digestibility, not the method of pulping. Kamstra et al. (1980) reported that ponderosa pine wood had a 60 percent in vitro digestibility after removal of 78 percent of the lignin with peroxyacetic acid.

A similar relationship was encountered with respect to the growth of the fungus *Aspergillus fumigatus* on a variety of commercial pulps prepared

TABLE 15 Degree of Delignification Required to Attain 60 Percent In Vitro Digestibility

Wood	Required Delignification[a] (%)	Lignin in Original Wood (%)
White birch	25	20
Red oak	35	23
Red pine	65	27
Douglas fir	73	32

[a] Based on original wood.

SOURCE: Baker (1973). Courtesy of *Journal of Animal Science*.

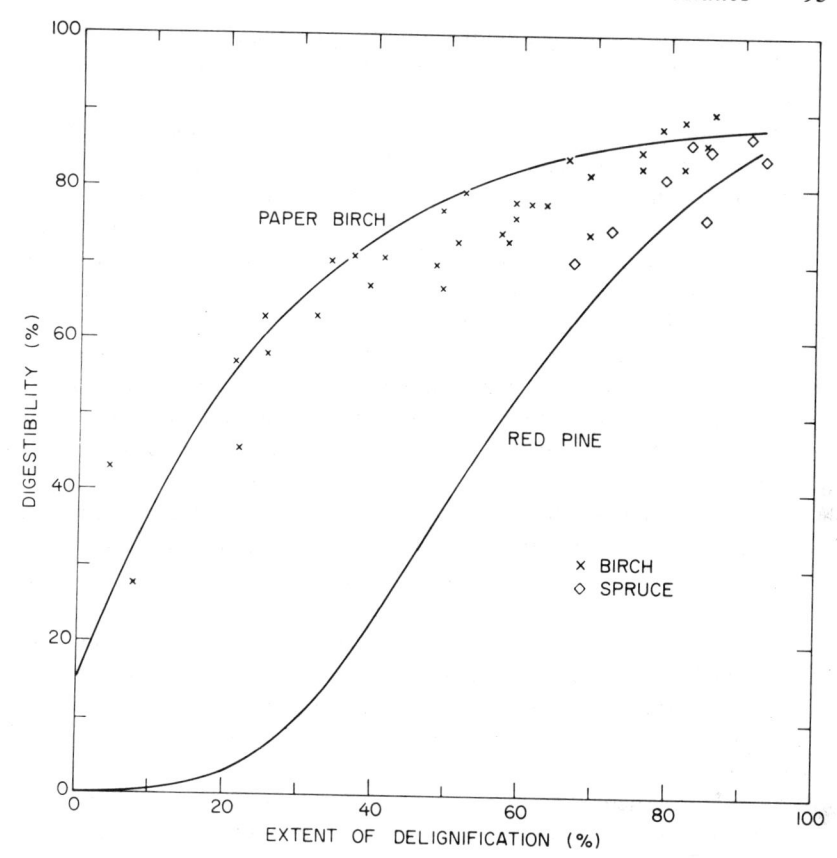

FIGURE 7 Relationship between digestibility and extent of delignification for wood pulps (data points from Saarinen et al., 1959; curves from Figure 6). SOURCE: Baker (1973).

under different conditions (Baker et al., 1973). As determined by the protein content of the fungal mass, reasonable growth on hardwood could be obtained at lignin contents of 14 percent or less, whereas fungal growth on softwoods was restricted to pulps having less than 3 percent residual lignin.

Because complete delignification is not required to prepare a nutritionally acceptable feedstuff, attention has turned to chemical pretreatments that can substantially disrupt the lignin-carbohydrate complex, but may demand less costly procedures than conventional pulping processes. Treatment with sodium chlorite has been one approach. In acetic acid solution, sodium chlorite exerts a marked specificity for lignin solubilization and has long been a quantitative technique for the preparation of "holocellulose," the total carbohydrate portion of a lignocellulose (Green, 1963). The high digestibility of a birch holocellulose was described by Saarinen et al. (1959).

All pulping procedures investigated thus far depend upon the selective removal of lignin for their beneficial effects. There is yet another possibility for enhancing the availability of wood-residue carbohydrates, that of disrupting the lignin-carbohydrate association in situ, without the selective removal of either constituent. Under proper conditions, it was found that gaseous sulfur dioxide can effect a disruption; the treatment appears applicable to both hardwoods and softwoods.

For this work, wood residues in the form of sawdust were reacted for 2 hours (hardwoods) or 3 hours (softwoods) at 120°C with an initial SO_2 pressure at room temperature of 207 kPa and a water-to-wood ratio of 3:1 (no free liquid) (Baker et al., 1975). After blowdown and a brief evacuation, the treated woods were neutralized to about pH 7 with sodium hydroxide and then air dried. Table 16 presents analytical data and values for 48-hour enzyme digestion for both the original woods and the treated products. A sample of alfalfa was included for comparison. The enzyme digestion was by the method of Moore et al. (1972).

As expected, cellulase digestion of the original woods was minimal, from a high of 9 percent for aspen to essentially 0 percent for the two softwoods. Even with alfalfa, only about half of the available carbohydrate was converted to sugars. Yields of the sulfur dioxide-treated products were 106 to 112 percent, based on starting material, as a result of the sulfonation and neutralization add-on reactions. Although all of the lignin was retained in the products, Klason lignin analysis of the five treated hardwoods showed lignin values of only 5 to 9 percent. This suggested

TABLE 16 Composition and Cellulase Digestion of Various Woods Before and After SO_2 Treatment

Wood	Lignin (%) Before	Lignin (%) After	Carbohydrate (%) Before	Carbohydrate (%) After	Digestibility (%) Before	Digestibility (%) After
Quaking aspen	20	7	70	71	9	63
Yellow birch	23	9	66	67	4	65
Sweetgum	20	5	66	64	2	67
Red oak	26	8	62	60	1	60
Douglas fir	30	24	65	63	0	46
Ponderosa pine	31	19	59	58	0	50
Alfalfa	17	—	51	—	25	—

SOURCE: Baker et al. (1975).

that the original lignin had been extensively depolymerized during sulfur dioxide treatment and converted to soluble lignins, a fact subsequently confirmed by extraction with boiling water. Depolymerization was less extensive with the two softwoods, and the higher Klason lignin values are reflected by a decreased digestibility. Enzymatic hydrolysis of the hardwood carbohydrates was essentially quantitative, indicating a complete disruption of the strong lignin-carbohydrate association in the original woods. The 60 to 65 percent digestibility of the treated hardwoods ranks these materials equivalent to moderate- or high-quality forage. The two softwood products would be equivalent to a low-quality forage, but might be upgraded through a better choice of processing conditions.

For information on palatability, possible toxic side effects, and in vivo nutritional value, a 136 kg batch of sulfur dioxide-treated material was prepared from red oak sawdust and fed to goats at levels of 0 to 50 percent of a pelleted roughage-type ration over an 8-week period. Average in vivo digestibilities for dry matter and carbohydrate as a function of wood content of the rations are plotted in Figure 8. Extrapolation of the curves to 100 percent sulfur dioxide-treated wood yielded values of about 52 percent for dry-matter digestion and 60 percent for carbohydrate digestion. From the shallow slope of the curves, it appears that a vapor-phase treatment with sulfur dioxide effectively converts red oak sawdust into a ruminant feedstuff having the dietary energy equivalence of a medium-quality forage. Neutralization of the treated product with ammonia rather than sodium hydroxide would augment its crude protein content.

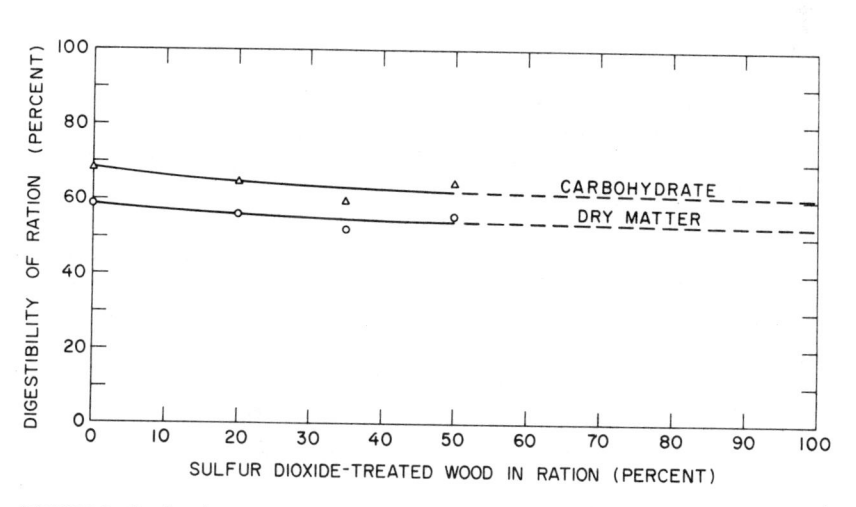

FIGURE 8 In vivo dry-matter digestion of rations containing SO_2-treated red oak. SOURCE: Baker et al. (1975).

Biochemical

Nine white-rot fungi were examined for their ability to remove lignin faster than they removed polysaccharides from aspen and from birch (Kirk and Moore, 1972). During decay most of the fungi decreased the lignin content of the aspen and the birch; that is, they removed a larger percentage of the lignin than of the polysaccharides. Lignin removal was always accompanied by removal of polysaccharides, but lignin removal did not correlate with removal of any particular component of the polysaccharides. During decay, lignin was usually more selectively removed in the first few percentage points of weight loss than were the polysaccharides. The decayed woods with less lignin were more digestible by rumen fluid than were the control samples. A summary of results of the study appears in Table 17. There is a very high inverse correlation between lignin content of the decayed wood and in vitro rumen digestibility. The coefficient of correlation is 0.95 for aspen and 0.97 for birch.

The results show that some white-rot fungi are effective in removing lignin faster than polysaccharides from wood. However, delignifying wood by these fungi under the test conditions is relatively slow. Optimizing factors such as aeration, moisture, temperature, and source and amount of nutrient nitrogen could enhance the rate of decay. In addition, it may be possible to find conditions that would improve the selectivity of lignin removal. It may also be possible to find better fungus species or better isolates than were used, or to obtain desirable mutants.

Grinding

Millett et al. (1970) found that, while vibratory ball milling is indeed an effective pretreatment, the milling response is quite species-selective. As an example, Figure 9 shows the in vitro rumen digestibilities of five hardwoods as a function of the time of milling. With 140 minutes of milling, all woods more or less attained a digestibility plateau. However, the plateau of digestibility was widely different for the different species, ranging from about 80 percent for aspen and sweetgum (*Liquidamber styraciflua*) to only 20 percent for red alder (*Alnus rubra*). Softwoods were even less responsive than red alder; five different softwoods showed a maximum digestion of only 18 percent after 120 minutes of vibratory milling. Thus, this selective species response severely limits the broad application of the milling technique. Time and energy costs impose further restrictions.

TABLE 17 Lignin and Carbohydrate Content and Digestibility of Sound and Decayed Aspen and Birch Wood

Fungus	Approximate Decay Time (days)	Average Weight Loss (%)	Range of Weight Losses (%)	Lignin (%)	Total Carbohydrates (%)	Rumen Fluid Digestibility (%)
Aspen						
None						
(control)	0	0	—	15.9	71.3	46
Fomes						
ulmarius	77	13	10–17	11.9	73.6	64
	71	51	46–56	18.7	61.5	46
Polyporus						
berkeleyi	88	21	16–23	8.5	70.5	77
Polyporus						
frondosus	64	20	19–22	10.5	71.8	71
	101	50	45–53	12.8	65.0	66
Polyporus						
giganteus	28	15	14–18	14.9	68.8	57
Polyporus						
versicolor	46	40	—	15.4	70.6	52
Birch						
None						
(control)	0	0	—	21.2	66.8	20
Fomes						
ulmarius	64	15	14–16	14.7	69.8	57
Ganoderma						
applanatum	34	17	15–20	21.1	63.4	29
	64	32	30–33	20.0	63.0	41
Polyporus						
berkeleyi	44	8	6–10	16.2	66.4	54
	69	22	19–26	16.0	64.1	60
Polyporus						
frondosus	44	14	12–16	14.7	66.8	63
Polyporus						
resinosus	40	11	9–14	15.7	65.7	57
	111	22	18–24	15.4	64.6	57

SOURCE: Kirk and Moore (1972). Courtesy of *Wood and Fiber Science*.

Irradiation

The technique of irradiating wood or straw by gamma rays or by high-velocity electrons substantially improves digestibility by rumen organisms (Lawton et al., 1951; Millett et al., 1970; Pritchard et al., 1962). However a strong species specificity appears again (see Table 18). The digestion of aspen carbohydrates is essentially quantitative after an electron dosage of 10^8 rad, while spruce is only 14 percent digestible at this dosage.

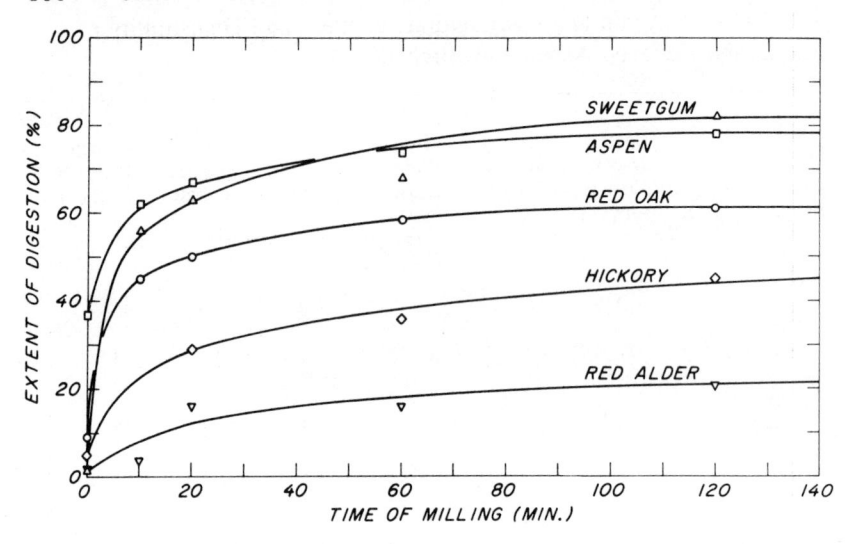

FIGURE 9 Relationship between in vitro digestibility and time of vibratory ball milling. SOURCE: Baker et al. (1975).

Utilization Systems

Experimental

Previous research has demonstrated that certain fractions of forest biomass can be made useful in animal diets without additional treatments. The research has also shown that many species and fractions of forest biomass are not useful and that physical and chemical treatments are needed to enhance digestibility. Advancement of the experimental systems requires

TABLE 18 Effect of Electron Irradiation on the In Vitro Digestion of Aspen and Spruce

	Digestibility (%)	
Electron Dosage (rads)	Aspen	Spruce
0	55	3
10^6	52	3
10^7	59	5
5×10^7	70	8
10^8	78	14

SOURCE: Baker et al. (1975).

a close and long-range working arrangement between animal and wood scientists.

Industrial

The use of products from wood for animal diets is currently in practice. These systems have developed slowly, perhaps because of the lack of working arrangements between animal scientists and the wood industry. Again, close research and working arrangements are needed to promote these systems. Closer cooperation should be sought between the feed industry and the wood industry to increase research on products from the forest biomass that can be useful in animal diets.

Potential

The forest biomass is the world's largest storehouse of carbohydrates. In the past it has been used in animal diets on a small scale in emergencies. There will be competition for the forest biomass, both for traditional forest products and, increasingly, for fuel. There will be very little in the way of unused forest products industry residues. By choosing proper growing, harvesting, and processing methods, forest biomass can be wisely used for traditional forest products, fuel, and livestock feed.

PULPMILL AND PAPERMILL RESIDUES

Quantity

The quantity, location, and composition of pulpmill and papermill primary sludges have been reported by Joyce et al. (1979). About one-third of the U.S. production of pulp and paper was included in the survey. The data are reported on geographic regions of the United States based on similarity of tree species, pulping processes, and end products within the region (see Figure 10). Table 19 shows the quantity of primary sludge by pulping process. These data indicate a wide range in rates of primary sludge production. The amount of sludge produced is related to the quantity of fines that can be tolerated in the final product and, in the case of de-inking, the quantity of fines that are in the raw material or developed in refining the raw material.

Table 20 shows the quantity of primary sludge produced per unit of production by the reporting mills. These data indicate an average of 84 kg/ton of production and a range of 2 to 400 kg/ton.

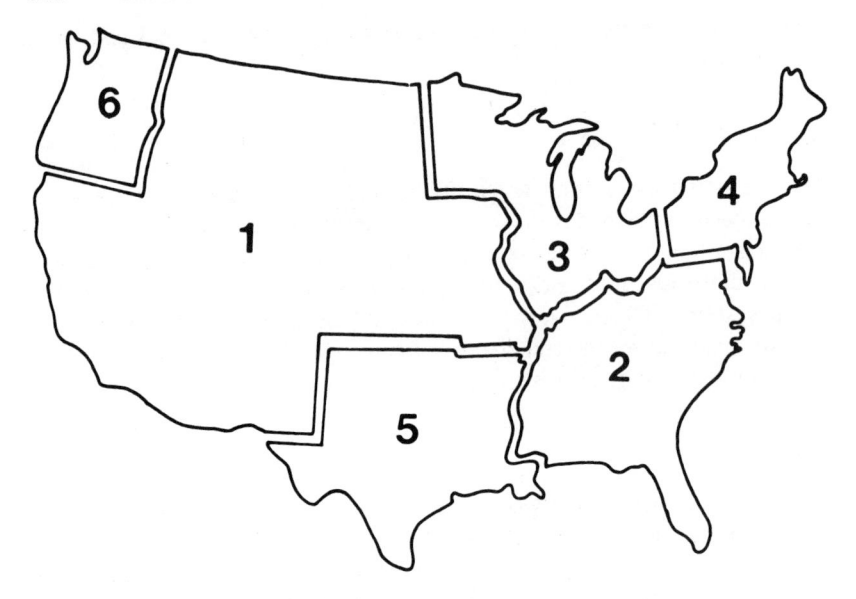

FIGURE 10 Division of the United States into six regions, based on best judgment of similar tree species, pulping processes, and end products within a region. SOURCE: Joyce et al. (1979). Courtesy of Elsevier, Amsterdam.

Additional data are reported from papermills (see Table 21). Some papermills make pulp by one process and purchase market pulp produced by another process.

Table 22 indicates the quantity of primary sludge production per unit of paper produced by region. This indicates an average of 52 kg/ton of paper production and a range of 2 to 185 kg/ton.

The type of mill and of product produced influences the composition of the primary sludge, as shown in Table 23. These data indicate that the sludge can contain 10 to 90 percent inorganic content.

The primary sludge data indicate that a substantial amount of cellulosic fiber is lost in pulp and papermaking processes. It is difficult to predict, however, how much usable fiber can be reclaimed from the primary sludges because of inherent variability of the manufacturing processes. The quantity and quality of the primary sludge are site-specific, and complete nutritional and chemical analyses are required to determine the usefulness of a sludge in animal diets. For use in animal diets, the fibers should be collected prior to entry into the primary collection site.

If the average primary sludge produced per ton of pulp is 84 kg and 50 percent of this is considered to be cellulosic fiber, at the 1980 pulp

TABLE 19 Kilograms Primary Sludge per 1000 Kilograms Pulp Produced by Different Pulping Processes

Pulp Process	Mills Reporting	Sludge Production Average	Minimum	Maximum
Soda	1	14	—	—
NSSC[a]	1	20	—	—
Kraft	27	58	17	103
Sulfite	3	102	22	234
Chemi-mechanical	4	113	40	204
De-inking	3	234	24	400

[a]Neutral sulfite semichemical pulping.

SOURCE: Joyce et al. (1979). Courtesy of Elsevier, Amsterdam.

production rate of nearly 52×10^6 tons/year, 2.2×10^6 tons/year of cellulosic fibers are available in the primary sludge.

North American sulfite pulp production and estimated associated sulfite spent liquor (SSL) solids appear in Table 24 (J. N. McGovern, University of Wisconsin, Madison, 1979, personal communication).

Technological advancements and economic and environmental demands have resulted in important changes in sulfite mill practice over recent years that have affected the nature, production, and availability of SSL solids. These changes have primarily involved substitution of the classical sulfite-pulping reagent of calcium bisulfite (in sulfurous acid) with the more soluble and recoverable magnesium, sodium, and ammonium compounds

TABLE 20 Kilograms Primary Sludge per 1000 Kilograms Pulp Production, by Region

Region[a]	Mills Reporting	Sludge Production Average	Minimum	Maximum
1	4	96	17	154
2	18	54	2	267
3	10	178	24	400
4	7	64	14	167
5	1	26	—	—
6	5	36	20	60
All reporting mills	45	84	2	400

[a]See Figure 10 for regions.

SOURCE: Joyce et al. (1979). Courtesy of Elsevier, Amsterdam.

TABLE 21 Kilograms Primary Sludge Produced per 1000 Kilograms Paper Produced by Different Types of Mills

Type of Mill	Mills Reporting	Sludge Production Average	Minimum	Maximum
NSSC[a]	2	12	8	17
Soda	1	15	—	—
Kraft	26	41	2	123
Chemi-mechanical	3	60	50	80
Sulfite	2	65	50	81
De-inking	4	105	24	185

[a]Neutral sulfite semichemical pulping.

SOURCE: Joyce et al. (1979). Courtesy of Elsevier, Amsterdam.

and recovery of the heat and chemicals in the SSL by evaporation, combustion, and chemical recovery. Evaporation and recovery of lignosulfonate by-products with prior desugaring to produce alcohol and yeast is practiced in several plants. A summary of these practices in U.S. sulfite mills is shown in Table 25 (J. N. McGovern, University of Wisconsin, Madison, 1979, personal communication).

There appears to be only one sulfite mill in the United States that does not recover its spent sulfite liquors, while 14 others evaporate and burn the liquor either to recover the pulping chemical base and sulfur dioxide or, in the case of ammonia-base pulping, to incinerate the spent liquor to recover the sulfur dioxide.

TABLE 22 Kilograms Primary Sludge Produced per 1000 Kilograms Paper Production, by Region

Region[a]	Mills Reporting	Sludge Production Average	Minimum	Maximum
1	4	30	15	50
2	17	43	2	114
3	15	79	8	185
4	9	41	15	100
5	1	23	—	—
6	3	44	20	62
All reporting mills	49	52	2	185

[a]See Figure 10 for regions.

SOURCE: Joyce et al. (1979). Courtesy of Elsevier, Amsterdam.

TABLE 23 Inorganic Content of Primary Sludges

Pulp Process	Mills Reporting Following Inorganic Content (%)					
	<10	11–20	21–40	41–60	61–80	>81
Kraft	5	4	7	9	2	0
Chemi-mechanical	1	0	0	0	0	0
NSSC[a]	0	1	0	0	0	1
Sulfite	2	0	0	0	0	0
Soda	0	0	1	0	0	0
De-inking	1	1	0	2	0	0
Total	9	6	8	11	2	1

[a]Neutral sulfite semichemical pulping.

SOURCE: Joyce et al. (1979). Courtesy of Elsevier, Amsterdam.

The SSL situation is quite different in Canada (see Table 26) where over 80 percent of the mills do not recover SSL solids and over half of the mills produce a high-yield newsprint pulp using a sodium acid sulfite process (J. N. McGovern, University of Wisconsin, Madison, 1979, personal communication). Two mills produce SSL by-products, but about 2.7 million tons of SSL solids are discharged unused.

TABLE 24 Sulfite Pulp and Spent Liquor Solids Production from North American Sulfite Mills

Pulp Grade	Number of Mills	Production (1,000 tons)	Spent Liquor Solids (1,000 tons)
United States	28[a]		
Unbleached		359	389
Bleached		1,653	1,836
Dissolving and high alpha		950[b]	952
Total		2,992	3,177
Canada	37		
Unbleached		2,352	3,352
Bleached		889	1,182
Dissolving and high alpha		378	580
Total		3,619	4,114

[a]27 as of 1979.
[b]Kraft excluded.

SOURCE: J. N. McGovern, University of Wisconsin, Madison, 1979, personal communication.

TABLE 25 Spent Sulfite Liquor Handling in United States Sulfite Mills (number of mills)

Reagent	Evaporation Combustion	Desugaring, Evaporation, By-Product Recovery	No Recovery	Total
Calcium	1	6	—	7
Magnesium	10	—	—	10
Sodium	—	1	—	1
Ammonia	3	5	1	9
Total	14	12	1	27

SOURCE: J. N. McGovern, University of Wisconsin, Madison, 1979, personal communication.

Physical Characteristics

Primary sludge or cleaner cellulosic fibers at pulpmills and papermills are collected wet from a water slurry. Primary sludges contain the material removed from the bottom of a primary collection pond. Primary sludge contains about 65 percent water and, due to its consistency, it is difficult to remove additional water mechanically. The cellulosic fibers can either be in the form of fine fibers too short for use in paper or in the form of screener or knotter rejects that are too large to be used directly. Most mills recycle the screener rejects to refiners, but the smaller mills often do not recycle them for economic reasons. The fiber fines are also difficult to dewater to less than 65 percent water content. The screener rejects can be squeezed to remove water to about 55 percent water content.

TABLE 26 Canadian Sulfite Pulpmills, by Process and Spent Liquor Solids Handling

Base	Mills Without Recovery Number	Pulp Capacity (tons/day)	Estimated Solids (tons/day)	Mills with Recovery Number	Pulp Capacity (tons/day)
Calcium	7	1,697	2,100	—	—
Magnesium	2	262	290	2	790
Sodium	19	6,850	6,850	2	360
Ammonia	3	1,275	1,275	2	1,200
Total	31	10,084	10,515	6	2,350

SOURCE: J. N. McGovern, University of Wisconsin, Madison, 1979, personal communication.

The fiber fines are usually broken cellulosic fibers and parenchyma cells that are less than 1 mm long and have passed through a 100-mesh screen. The screener rejects are undercooked wood that sometimes has come from a knot in the tree, hence the name knotter rejects. These rejects are from 20 to 50 mm long and contain delignified wood on the surface and nearly whole wood on the inside. The primary sludge can contain these materials and, in addition, wood and bark fines from the wood room, dirt, grit, and papermaking additives such as coating clay, calcium carbonate, and titanium dioxide. The inorganic content is usually in the form of less than 200-mesh fines.

Nutritive Value

Fines and Screen Rejects

Four fiber types constitute the bulk of residues generated by pulpmills: (1) groundwood fines, fully lignified fiber fragments created during grinding or milling operations in newsprint mills; (2) semichemical pulping fines, partially delignified fiber fragments produced in the manufacture of pulps for corrugating materials; (3) screen rejects, which are partially pulped but contain unbleached chip fragments and fiber bundles that are undesired by-products of all chemical pulping operations; and (4) chemical pulping fines, fiber fragments generated in all pulp-making procedures along with ray parenchyma cells and vessel elements from hardwood tissue mill operations.

Chemical Composition Information on the composition and in vitro rumen digestibility of several pulpmill residues is shown in Table 27 and in Appendix Tables 1 to 4. Fines and screen rejects are very low in protein, and high in acid detergent fiber. Their mineral content reflects the exposure to chemical processing. Care must be used in feeding some of these residues because potentially toxic levels of mineral elements could be present, depending upon how the papermill handles its waste streams.

Nutrient Utilization Digestibility values in Table 27 are expressed in terms of percentage weight loss after 5 days of incubation with buffered rumen fluid at 39°C (Mellenberger et al., 1970). As expected, groundwood fines yielded digestibility values comparable to those observed for sawdust of the same species: 0 percent for the pine and spruce and about 35 percent for aspen. All of the screen rejects and chemical pulp fines had digestibilities of more than 40 percent, and digestibility of two of the pulp fines was more than 70 percent. Thus, based on in vitro dry-matter digestibility,

TABLE 27 Composition and In Vitro Rumen Digestibility of Pulpmill Residues

Type of Residue	Lignin	Carbohydrate	Ash	In Vitro Digestibility (%)
Groundwood fines				
Aspen	21	73	1	37
Southern pine	31	59	1	0
Spruce	31	60	1	0
Screen rejects				
Aspen sulfite	19	77	2	66
Mixed hardwood, sulfite	24	65	14	54
Mixed hardwood, kraft	25	74	9	44
Chemical pulp fines				
Mixed hardwood, kraft (bleached)	< 1	109	1	95
Aspen sulfite (parenchyma cells)	20	73	2	73
Southern pine, kraft (unbleached)	28	68	4	46

SOURCE: Baker et al. (1975).

any of the screen rejects and chemical pulp fines could serve as useful sources of dietary energy for ruminants. The mixed hardwood, bleached-kraft chemical pulp fines are essentially pure cellulose. Table 27 shows that the lignin and total carbohydrate contents of the aspen groundwood, aspen sulfite screen rejects, and aspen sulfite parenchyma cell fines are almost identical, whereas the in vitro dry-matter digestibility ranges from 37 to 73 percent. The digestibility of fines of aspen parenchyma cells, for example, is higher than would be predicted on the basis of lignin content because the parenchyma cells contain substances that analyze as Klason lignin. Southern pine kraft (unbleached) pulp also contains substances that could analyze as lignin.

In general, animals have readily accepted diets containing up to 75 percent pulp residues, and digestibility of the residues has been quite acceptable. For example, Millett et al. (1973) determined the digestibility of pelleted diets containing increasing levels of three types of pulpmill residues up to 50 percent of the total diet. Estimates of residue digestibility were obtained by regressing digestibility on the amount of residue in the diet. Mixed hardwood bleached-kraft fines were estimated to be 80 percent digestible, while screen rejects were 66 percent and parenchyma fines 50 percent. In vitro digestibilities of the three residues were 95, 66, and 73 percent, respectively. The relatively low in vivo utilization of the carbohydrates in the aspen sulfite parenchyma fines may be related to their small particle size (about 0.1 mm long), which might favor short retention times and inadequate bacterial action in the rumen.

Additional confirmation of the feeding value of wood pulps is provided by Saarinen et al. (1959). In vivo digestibilities of wood pulps prepared by various pulping techniques ranged from 27 to 90 percent. While these were experimental pulps, the results indicate the digestibility range to be expected from the fines fractions generated by commercial pulping operations.

Dinius and Bond (1975) fed commercially available unbleached fiber fines along with activated sludges as a protein source to steers. The fines were screened from maple, birch, and beech pulp made by an ammonia-base sulfite pulping process. The effects of feeding pulp fines on ruminal pH, ammonia, and volatile fatty acid concentrations and on voluntary intake and growth of pregnant beef heifers throughout gestation were also measured. The fines were estimated to be 92.8 percent digestible. Weight gain, calf birth weight, or incidence of calving problems for heifers fed diets containing 75 percent fines did not differ from those of heifers fed hay. Lemieux and Wilson (1979) used pulp fines from the same commercial source and noted that diets containing 20 to 67 percent of the wood residue were as digestible as the corn-hay control diets. The growth rate of lambs fed these diets was comparable to that of the control lambs. There were no adverse effects of wood-containing diets on blood serum urea, protein or mineral levels, or rumen and liver histopathology.

Animal Performance Millett et al. (1973) offered Hereford steers diets containing 0 and 50 percent unbleached southern pine kraft pulp fines for a period of 58 days. Average daily gain was 0.77 and 0.54 kg for the 0 and 50 percent pulp diets, respectively. Daily feed intake as a percentage of body weight was 2.84 and 2.48, and feed/gain ratio was 9.4 and 11.7, respectively. Steers fed the pulp diet tended to sort the ingredients and reject the fines fraction, indicating a lower palatability for this material.

Fritschel et al. (1976) conducted five different feeding experiments with sheep and steers in which aspen sulfite parenchyma fines were fed at levels of 72 and 83 percent of the total diet. One experiment included pregnant ewes fed over an 11-month period, including lambing and lactation. Results indicated ready acceptance of the diet by the ewes. The number and growth of lambs was equivalent to those from control hay-fed ewes. Equally good results were obtained with a group of Angus beef cows, some of which were pregnant at the beginning of the experiment and calved during the trial.

The value of pulp-containing diets in lamb-fattening operations was tested in two feeding experiments by Riquelme et al. (1975). In one, a diet of 60 percent alfalfa and 40 percent concentrate was compared with two diets containing 66 percent pulp fibers (a bleached hardwood kraft and a bleached mixed species sulfite), 5 percent alfalfa, and 29 percent concentrate. In the second, a diet of 70 percent barley and 30 percent alfalfa was compared with those diets containing approximately 70 percent of the bleached sulfite pulp and three different nitrogen sources. In all cases the pulp-fed lambs had desirable carcass characteristics equivalent to or higher than the controls. Based on growth measurements, the pulp fiber was considered to have about 85 percent of the energy value of barley.

Processing

Many pulp and paper residues are highly digestible, and it is not expected that they will require processing. In the case of utilization of the fines from a mechanical pulpmill where the fines are merely groundwood, processing requirements are the same as for wood and sawdust of the same species. If the materials are too contaminated with inorganics or other papermaking additives, it is doubtful if further processing could be justified on the basis of economic and animal health considerations.

Utilization Systems

Presently, fines from a sulfite pulpmill are being incorporated into livestock diets. The fines contain parenchyma cells, short fibers, and fiber bundles obtained from a pulpmill making fiber for tissue from birch, beech, and maple species. The process is an ammonia-base sulfite pulping process. The experimental program started with the testing of fines from an identical pulpmill operated by the same company that made pulp from aspen. The testing consisted of chemical analysis of the fines for lignin, carbohydrate content, ash, mineral elements, fats, and in vitro rumen digestibility. In vivo digestibility was determined with goats, and additional feeding trials with sheep were made to determine palatability and the effect of palatability on general health indicators (Millett et al., 1973). The fines used are identified in Millett's report as aspen sulfite parenchyma cell fines (unbleached). Additional feeding trials were made with beef animals (Fritschel et al., 1976). Nearly 50,000 kg were fed during a 3-year cooperative study between the Forest Products Laboratory, University of Wisconsin-Madison, and Procter and Gamble Company. Additional feeding trials were conducted by the Agricultural Research Service (Dinius and Bond, 1975). As a result of this and other research (Lemieux and

Wilson, 1979), pulp fines containing about 65 percent water are presently being fed to dairy cows and beef animals in Pennsylvania.

This experimental program required close cooperation between researchers of various disciplines, pulpmill management personnel, and corporate research staff. At the end, it involved close cooperation between pulpmill management and county agricultural agents to explain to the farmers how the fines are produced and how they should be used.

The use of additional spent sulfite pulp fines in animal diets requires study to determine how to make the material available without causing serious changes in the material and energy balances in the pulpmills and how best to utilize it near its source.

Potential

The results of these studies involving pulpmill residues showed that the cellulosic fiber fines from two pulpmills, each producing about 31 metric tons (dry basis) per day, could be used in animal diets. The extension of these results to other pulpmill residues could have a lasting and continued impact on agricultural land use, food production, and wise use of our renewable resources.

Sludges

Chemical Composition

In addition to the residues described above, pulpmills and papermills may produce primary clarifier or lagoon sludges. Table 28 shows the composition and the in vitro dry-matter digestibility of various combined pulpmill and papermill sludges (Millett et al., 1973). Most of the sludge-containing by-products are very high in ash, and moderate to high in lignin. Both of these characteristics diminish the feeding value of the sludge. The Klason lignin values in Table 28 also include acid-insoluble paper additives (ash) as lignin. Errors in the lignin analysis are evident in the data listed in Table 21 for the combined pulpmill and papermill residues that have high ash content.

Nutrient Utilization

Because the groundwood mill sludges are mostly groundwood, digestibility is expected to be low, although the total carbohydrate content is high. One of the semichemical pulpmill sludges was high in digestibility

TABLE 28 Composition and In Vitro Rumen Digestibility of Combined Pulpmill and Papermill Sludges

Type of Residue	Klason[a] Lignin (%)	Total Carbohydrate (%)	Ash (%)	In Vitro Dry-Matter Digestibility (%)
Groundwood mill				
Mixed species plus some mixed chemical pulps	50	41	38	24
Southern pine plus some hardwood kraft	24	60	15	19
Semichemical pulpmill				
Aspen	20	71	2	57
Aspen plus mixed hardwoods	55	29	13	6
Chemical pulpmill				
De-inked wastepaper, tissue	23	71	22	72
Softwood sulfite, glassine	13	74	14	66
Reprocessed milk carton stock	28	67	25	65
Mixed chemical pulps, tissue	17	76	13	60
Mixed hardwood bleached kraft, printing	17	75	11	59
Aspen sulfite, tissue	19	77	2	50
Aspen and spruce sulfite, tissue	45	46	45	35
Secondary waste treatment sludge	38	5	45	5

[a]Includes ash not soluble in H_2SO_4.

SOURCE: Millett et al. (1973). Courtesy of *Journal of Animal Science*.

and total carbohydrate and low in ash, but the other was low in digestibility and total carbohydrate. This indicates the amount of variation that can be observed between products from mills that use the same pulping process.

Dinius and Bond (1975) fed a mixture of 65 percent pulp fines and 35 percent sludge. The activated sludge was the residue from processing total pulpmill wastes through a lagoon system and constituted 30 percent of the diet. At 38 percent protein, it provided the sole nitrogen source. The remaining 5 percent of the diet consisted of inorganic salts. Organic-matter digestibility of this diet was 78.9 percent, but animal acceptance of the diet was poor. Dry-matter consumption was only 1.2 percent of body weight.

Spent Sulfite Liquor

The chemical composition of spent sulfite liquor (SSL) varies greatly and depends upon wood species, pulping reagent, reagent composition, and pulping conditions. For a normal yield of calcium bisulfite-sulfurous acid reagent pulp, the spent sulfite liquor can contain 12 to 16 percent total solids, 30 percent of which consists of monosaccharides with some oligosaccharides. The sugars in the presently unutilized spent sulfite liquor can be concentrated and mixed with molasses. Klopfenstein et al. (1973) reported that calcium and ammonia spent sulfite liquor can be effectively used as a liquid feed ingredient in beef cattle diets, but that it contains only about 50 percent as much energy as cane molasses per unit of dry matter because of lignin residues.

WOOD RESIDUES AS ROUGHAGE SUBSTITUTES IN RUMINANT DIETS

Ruminants generally require a certain level of roughage in their diet. Roughage provides tactile stimulation of the rumen wall and promotes rumination, which in turn increases salivation and supply of buffer for maintenance of rumen pH. Roughage in the dairy diet can help maintain normal milk fat test, and in high-grain feedlot cattle diets can lower the incidence of rumen parakeratosis and liver abscess. When traditional sources of forage or roughage are expensive or in short supply, alternative roughage sources could be helpful. Wood residues have been investigated for this purpose.

Aspen sawdust has been shown effective as a partial forage substitute in a high-grain dairy diet (Satter et al., 1970). Cows fed 2.3 kg hay and about 17 kg of a pelleted diet containing one-third aspen sawdust maintained normal milk fat. Cows receiving a similar diet but without sawdust produced milk with 50 percent as much fat. Diets with one-third sawdust are not practical, however, for high-producing dairy cows. A subsequent study (Satter et al., 1973) was done in which lower levels of sawdust were fed with high-grain diets, with or without sodium bicarbonate and sodium bentonite. It was concluded that aspen sawdust can be a partial roughage substitute in lactating dairy cow diets and can be helpful in maintaining near-normal milk fat content in high concentrate diets. The sawdust cannot serve as the only source of roughage for lactating cows because of the irregular feed intake that results if no other forage is fed.

Oak sawdust is essentially indigestible, and has been used in several growth or feedlot trials as a roughage substitute (Anthony and Cunningham, 1968; Dinius et al., 1970; El-Sabban et al., 1971). Inclusion of 5 to 15 percent oak sawdust in the diet has generally supported animal performance equal to the other experimental groups being tested. Unfortunately, the design of these experiments or the numbers of animals used prevent firm conclusions regarding the value of oak sawdust as a direct replacement for conventional roughage. The study of El-Sabban et al. (1971) does suggest that the incidence of liver abscesses did decrease with increase in dietary sawdust up to 15 percent of the diet. Rumens of steers fed diets containing sawdust were parakeratotic, but did show improvement, particularly when coarse sawdust particles were fed. It has been generally concluded that oak sawdust is an effective roughage substitute when used as 5 to 15 percent of the total diet. Similar conclusions were reached when pine sawdust was used as a roughage substitute in beef finishing diets (Slyter and Kamstra, 1974).

ANIMAL HEALTH

Generally, whole-tree or tree residues are not considered dangerous to the health of livestock. It is essential, however, that diets containing wood residues be properly balanced for all of the essential nutrients. Wood residues must be considered primarily as energy sources. They contain only small amounts, or are nearly devoid, of many essential nutrients. Thus, animals consuming diets containing large amounts of wood residues will encounter ill health if the diets are improperly balanced.

Pine needles have been shown to be toxic and can cause abortion in cattle. Pregnant mice fed an extract of ponderosa pine (*Pinus ponderosa*) needles had reduced litter size, and thus can serve as indicators of potential toxicity problems (Cogswell and Kamstra, 1980).

Wood residues that have been exposed to chemical treatments, such as in the pulping process, should be carefully examined for residual chemicals. In the pulp and papermaking industry, the process or method of handling waste streams can unwittingly result in contamination of the potential feedstuff. Sludges in particular need careful scrutiny. A number of sludges have digestibility values comparable to hay. Their suitability for animal feed, however, will depend on the amount of ash and the chemical nature of the individual ash constituents. For example, moderate levels of clay-type filler could be tolerated, but the presence of more than trace amounts of certain heavy metals would rule out use as a feedstuff.

Thus, each pulp and papermaking residue should be chemically characterized before it can be recommended as a feedstuff.

REGULATORY ASPECTS

The use of wood and wood-derived products in animal feeds requires the consideration of and compliance with state and federal regulations.

The Association of American Feed Control Officials (1980) reported that ground whole-tree aspen (*Populus* spp.) and/or its parts can be fed to animals provided that the diets are supplemented with protein, vitamins, and minerals.

RESEARCH NEEDS

The relatively low digestibility of most wood residues has stimulated research into ways of treating or processing wood residues to increase digestibility. A number of approaches have proven effective in increasing digestibility, but they are costly and not economically feasible. The need for a cost-effective process is clear. While the need is very great for such a process, the likelihood of developing one may be diminished as cost of energy and chemical inputs increase.

A major portion of the total wood residue is the residue that is left in the forest at the time of harvest. This suggests that improved techniques for whole-tree harvest be developed so the heretofore underutilized portions of the tree may be available for collection, processing, and transport to a site for utilization. In this connection, more research is needed on the nutritive value of the foliage portion of trees used for pulpwood. Commercial equipment exists for chipping whole trees, and sorting out the wood chips from leaf, bark, and twig material at the point of harvest. As this method of harvesting grows, so will the potential supply of foliage material. Much more information is needed on the feeding value of the foliage, and how feeding value is influenced by time of harvest, storage of the foliage, and the genetic background of the tree (clone from which the foliage was derived).

Most of the wood residues that have potential as feedstuffs are relatively high in moisture. This causes problems in storage because of deterioration. Research is needed to identify low cost means of preserving high moisture wood residues. Preservation through ensiling and the use of mold inhibitors are approaches that might be fruitful.

A significant portion of the research that has been done with wood residues has involved sheep or goats. Unfortunately they are not as well suited as cattle are to digest low-quality fibrous feeds. Sheep and goats are selective browsers, and normally select the more easily digested plant material from their grazing environment. Cattle, because of their larger size and capacity for retaining slowly digested materials in the rumen for longer periods of time, can more easily accommodate the poorer-quality feedstuffs (Van Soest, 1982). Future research with the relatively low quality wood residues should, wherever possible, utilize cattle rather than sheep or goats.

SUMMARY

The carbohydrates of whole-wood residues are, with few exceptions, resistant to attack by cellulolytic organisms in the rumen. This resistance apparently stems from the close physical and chemical association between cellulose and lignin, augmented by the crystalline nature of cellulose itself. Whether or not the carbohydrates contained in wood lignocellulosic residues can be utilized by rumen microbes will depend largely on how extensively the lignin-carbohydrate complex can be altered or opened up.

Of the woods tested, all of the coniferous species are essentially undigested by rumen microorganisms. Deciduous species, with a few exceptions, are only slightly digested. Aspen is the most digestible species tested, giving both an in vitro and in vivo digestibility of about 35 percent. Aspen bark is about 50 percent digestible.

All of the chemical and physical pretreatments discussed are effective to some extent, but exhibit a strong species preference that severely limits their applicability. Hardwoods are generally more responsive to pretreatment than softwoods, but even hardwoods exhibit a broad range of responsiveness. Aspen is particularly susceptible to treatment. Several of the treatment methods, technically speaking, can be readily adapted to a commercial process. The cost of treatment has presented the biggest barrier. Conventional feedstuffs need to be relatively high priced before treated wood residues can compete in the marketplace. Prices of conventional feedstuffs have on occasion reached levels that would make treated wood residues attractive. The lack of a steady market has discouraged development of commercial wood processing enterprises.

Wood has been shown to be effective as a roughage replacement. Depending upon the other dietary ingredients, concentrations of 5 to 15 percent screened sawdust in diets for beef cattle appear practical. For lactating dairy cows, aspen sawdust could be used as a roughage extender or as a partial roughage substitute in high-grain diets. Some long hay appears to be necessary in the diet to stabilize feed intake.

Finally, it must be realized that wood residues are generally low in protein and other essential nutrients that livestock require. This necessitates more extensive supplementation. Treated wood residues are primarily an energy source, and some may compare as an energy source to an average- or low-quality hay. For this reason, treated residues are best suited for ruminants having relatively low nutrient requirements, such as overwintering beef cows and ewes and "dry" dairy cows, and for larger size dairy and beef replacement heifers. Foliage does, however, contain significant protein and other essential nutrients that could make foliage quite useful as a feedstuff. If the foliage can be economically harvested and stored, it may be a potential animal feedstuff.

Some of the many pulp and papermaking residues that are already partially delignified but that have little fiber value for paper manufacture have excellent potential as ruminant feedstuffs. Care must be used in their selection as feedstuffs, however, because some residues may contain toxic materials.

LITERATURE CITED

Anthony, W. B., and J. T. Cunningham, Jr. 1968. Hardwood sawdust in all concentrate rations for cattle. J. Anim. Sci. 27:1159 (Abstr.).

Apgar, W. P., H. C. Dickey, and H. E. Young. 1977. Estimated digestibility of conifer muka fed to sheep. Res. Life Sci., University of Maine at Orono 24:1.

Archibald, J. G. 1926. The composition, digestibility and feeding value of hydrolyzed sawdust. J. Dairy Sci. 9:257.

Association of American Feed Control Officials. 1980. Official publication of Association of American Feed Control Officials, Box 3160, College Station, Tex.

Baker, A. J. 1973. Effect of lignin on the in vitro digestibility of wood pulp. J. Anim. Sci. 36:768.

Baker, A. J., M. A. Millett, and L. D. Satter. 1975. Wood and wood-based residues in animal feeds. P. 75 in Cellulose Technology Research. A. F. Turbak, ed. ACS Symposium Series 10. Washington, D.C.: American Chemical Society.

Baker, A. J., A. A. Mohaupt, and D. F. Spino. 1973. Evaluating wood pulp as feedstuff for ruminants and substrate for *Aspergillus fumigatus*. J. Anim. Sci. 37:179.

Bender, F., D. P. Heaney, and A. Bowden. 1970. Potential of steamed wood as a feed for ruminants. For. Prod. J. 20:36.

Bender, R. 1979. Method of treating lignocellulose materials to produce ruminant feed. U.S. Pat. No. 4,136,207. Jan. 23.

Braconnot, H. 1819. Hydrolysis of cellulose into sugar. Gilbert's Annalen der Physik. 63:348.

Cogswell, C., and L. D. Kamstra. 1980. Toxic extracts in ponderosa pine needles. J. Range Manage. 33:46.

Dinius, D. A., and J. Bond. 1975. Digestibility, ruminal parameters and growth by cattle fed a waste wood pulp. J. Anim. Sci. 41:629.

Dinius, D. A., A. D. Peterson, T. A. Long, and B. R. Baumgardt. 1970. Intake and digestibility by sheep of rations containing various roughage substitutes. J. Anim. Sci. 30:309.

Edin, H. 1940. Experiments in the use of chemical pulp as fodder. Sven. Papperstidn. 43:82.

Edin, H., T. Helleday, and S. Nordfeldt. 1941. Feed cellulose for milk cows and horses, including other feeding experiments during the shortage and crisis in feedstuffs. Lantbrukshoegsk. Husdjoursfoersoeksanst. Medd. No. 6:1.

Ek, A. R., and D. H. Dawson. 1976. Actual and projected growth and yields of *Populus* "Tristis #1" under intensive culture. Can. J. For. Res. 6:132.

El-Sabban, F. F., T. A. Long, and B. R. Baumgardt. 1971. Utilization of oak sawdust as a roughage substitute in beef cattle finishing rations. J. Anim. Sci. 32:749.

Enzmann, J. W., R. D. Goodrich, and J. C. Meiske. 1969. Chemical composition and nutritive value of poplar bark. J. Anim. Sci. 29:653.

Feist, W. C., A. J. Baker, and H. Tarkow. 1970. Alkali requirements for improving digestibility of hardwoods by rumen micro-organisms. J. Anim. Sci. 30:832.

Fritschel, P. R., L. D. Satter, A. J. Baker, J. N. McGovern, R. J. Vatthauer, and M. A. Millett. 1976. Aspen bark and pulp residue for ruminant feedstuffs. J. Anim. Sci. 42:1513.

Gerry, R. W., and H. E. Young. 1977. A preliminary study of the value of conifer muka in a broiler ration. Res. Life Sci., University of Maine at Orono 24:1.

Gharib, F. H., R. D. Goodrich, J. C. Meiske, and A. M. El Serafy. 1975. Effects of grinding and sodium hydroxide treatment on poplar bark. J. Anim. Sci. 40:727.

Green, J. W. 1963. Methods in carbohydrate chemistry. P. 12 in Vol. III, Cellulose. New York: Academic Press.

Harris, E. E., G. J. Hajny, and M. C. Johnson. 1951. Protein evaluations of yeast grown on wood hydrolyzate. Ind. Eng. Chem. 43:1593.

Heaney, D. P., and F. Bender. 1970. The feeding value of steamed aspen for sheep. For. Prod. J. 20(9):98.

Hunt, J. R., and G. M. Barton. 1978. Nutritive value of spruce muka (foliage) for the growing chick. Anim. Feed Sci. Technol. 3:63.

Hvidsten, H. 1940. Fodder cellulose, its use and nutritive value. Nord. Jordbrugsforsk. 22:180. Biol. Abstr. 22(5):1123(1948).

Hvidsten, H., and T. Homb. 1951. Survey of cellulose and Beckmann-treated straw as feed. Pure Appl. Chem. 3:113.

Inman, R. E. 1977. P. 62 in Silvicultural Biomass Farms. Vol. I Summary. McLean, Va.: MITRE Corp.

Isebrands, J. G., J. A. Sturos, and J. B. Crist. 1979. Integrated utilization of biomass. A case study of short-rotation intensively cultured *Populus* raw material. Tappi 62(7):67.

Jelks, J. W. 1976. Process for oxidizing and hydrolyzing plant organic matter particles to increase the digestibility thereof by ruminants. U.S. Pat. No. 3,939,286. Feb. 17.

Joyce, T. W., A. A. Webb, and H. S. Dugal. 1979. Quantity and composition of pulp and papermill primary sludges. Resour. Recovery Conserv. 4:99.

Kamstra, L. D., D. Ronning, H. G. Walker, G. O. Kohler, and O. Wayman. 1980. Delignification of fibrous wastes by peroxyacetic acid treatments. J. Anim. Sci. 50:153.

Keays, J. L. 1976. Foliage. 1. Practical utilization of foliage. Appl. Polymer Symp. 28:445.

Keays, J. L., and G. M. Barton. 1975. Recent Advances in Foliage Utilization. Information Report VP-X-137. Vancouver: Canadian Forestry Service, Forest Products Laboratory.

Kirk, T. K., and W. E. Moore. 1972. Removing lignin from wood with white-rot fungi and digestibility of resulting wood. Wood and Wood Fiber 4(2):72.

Klopfenstein, T., W. Koers, and S. Farlin. 1973. Sulfite liquor in beef cattle rations. Pp. 25–26 in 1973 Nebraska Beef Cattle Report. ED 73-218. Omaha: University of Nebraska.

Kressman, F. W. 1922. The manufacture of ethyl alcohol from wood waste. USDA Bull. No. 983.

Lawton, E. J., W. D. Bellamy, R. E. Hungate, M. P. Bryant, and E. Hall. 1951. Some effects of high velocity electrons on wood. Science 113:380.

Lemieux, P. G., and L. L. Wilson. 1979. Nutritive evaluation of a waste wood pulp in diets for finishing lambs. J. Anim. Sci. 49:342.

Lloyd, R. A., and J. F. Harris. 1955. Wood hydrolysis for sugar production. Report No. 2029. Madison, Wis.: USDA Forest Products Laboratory.

Marchessault, R. H., and J. St. Pierre. 1978. A new understanding of the carbohydrate system. Chemrawn Conf., Toronto, Ontario.

Mellenberger, R. W., L. D. Satter, M. A. Millett, and A. J. Baker. 1970. An in vitro technique for estimating digestibility of treated and untreated wood. J. Anim. Sci. 30:1005.

Mellenberger, R. W., L. D. Satter, M. A. Millett, and A. J. Baker. 1971. Digestion of aspen, alkali-treated aspen and aspen bark by goats. J. Anim. Sci. 32:756.

Millett, M. A., A. J. Baker, W. C. Feist, R. W. Mellenberger, and L. D. Satter. 1970. Modifying wood to increase its in vitro digestibility. J. Anim. Sci. 31:781.

Millett, M. A., A. J. Baker, L. D. Satter, J. N. McGovern, and D. A. Dinius. 1973. Pulp and papermaking residues as feedstuffs for ruminants. J. Anim. Sci. 37:599.

Monteith, D. B. 1979. Whole tree weight tables for New York. Appl. For. Res. Inst. Res. Rep. No. 40. Syracuse: State University of New York.

Moore, W. E., M. J. Effland, and M. A. Millett. 1972. Hydrolysis of wood and cellulose with cellulytic enzymes. J. Agric. Food Chem. 20:1173.

Morrison, F. B., G. C. Humphrey, and R. S. Hulce. 1922. Unpublished Report. Madison, Wis.: USDA Forest Products Laboratory.

Nehring, K., and J. Schütte. 1950. Composition and nutritive value of foliage and twigs. I. The change in composition of leaves and twigs from different seasons. Arch. Tierernahr. 1:151.

Nehring, K., and W. Schramm. 1951a. Composition and nutritive value of leaves and branches. II. The digestibility of summer leaves and twigs. Arch. Tierernahr. 2:264.

Nehring, K., and W. Schramm. 1951b. Composition and nutritive value of leaves and branches. III. Nutritive value of dead leaves and winter branches. Arch. Tierernahr. 6:342.

Nordfeldt, S. 1951. Problems in animal nutrition. The use of wood pulp for feeding farm animals; the value of silage made with added acids. Proc. 11th Inst. Cong., Pure and Appl. Chem. 3:391.

Pew, J. C., and P. Weyna. 1962. Fine grinding, enzyme digestion and the lignin-cellulose bond in wood. Tappi 45:247.

Pritchard, G. I., W. J. Pigden, and D. J. Minson. 1962. Effect of gamma radiation on the utilization of wheat straw by rumen micro-organisms. Can. J. Anim. Sci. 42:215.

Ribe, J. H. 1973. Puckerbush weight tables. Life Sci. and Agric. Exp. Stn. Misc. Rep. 152. Orono: University of Maine.

Riquelme, E., I. A. Dyer, L. E. Bariba, and B. Y. Couch. 1975. Wood cellulose as an energy source in lamb fattening rations. J. Anim. Sci. 40:977.

Robertson, J. A., S. E. Beacom, and R. Shiels. 1971. Feeding value of poplar silage in rations for yearling steers. Can. J. Anim. Sci. 51:243.

Saarinen, P., W. Jenson, and J. Alhojärri. 1958. Investigation on cellulose fodder. I. Birch wood. Pap. Puu 40:495.

Saarinen, P., W. Jenson, and J. Alhojärri. 1959. Digestibility of high yield chemical pulp and its evaluation. Acta Agral. Fennica 94:41.

Satter, L. D., A. J. Baker, and M. A. Millett. 1970. Aspen sawdust as a partial roughage substitute in a high-concentrate dairy ration. J. Dairy Sci. 53:1455.

Satter, L. D., R. L. Lang, A. J. Baker, and M. A. Millett. 1973. Value of aspen sawdust as a roughage replacement in high-concentrate dairy rations. J. Dairy Sci. 56:1291.

Scott, R. W., M. A. Millett, and G. J. Hajny. 1969. Wood wastes for animal feeding. For. Prod. J. 19(4):14.
Segel, L., L. Loeb, and J. J. Creely. 1954. An X-ray study of the decomposition product of the ethylamine-cellulose complex. J. Poly. Sci. 13:193.
Sherrard, E. C., and G. W. Blanco. 1921. Preparation and analysis of a cattle food consisting of hydrolyzed sawdust. J. Ind. Eng. Chem. 13:61.
Simonsen, E. 1898. Vorläufige resultate der Fabrikmässigen versuche mit darstellung von Spiritus aus Sägespähner. Zeit. Angew. Chem. Part 1:962. Part II:1007.
Singh, M. 1978. Utilization of whole aspen tree material as a ruminant feed component. Ph.D. Thesis. South Dakota State University.
Singh, M., and L. D. Kamstra. 1981a. Utilization of whole aspen tree materials as a roughage component in growing cattle diets. J. Anim. Sci. 53:551.
Singh, M., and L. D. Kamstra. 1981b. Utilization of aspen trees as a ruminant feed component. Proc. S.D. Acad. Sci. 60:54.
Singh, M., L. D. Kamstra, J. A. Minyard, D. E. Moore, and R. Healy. 1978. Winter feeding demonstration with pregnant stock cows. Proc. S.D. Acad. Sci. 57:46.
Slyter, A. L., and L. D. Kamstra. 1974. Utilization of pine sawdust as a roughage substitute in beef finishing rations. J. Anim. Sci. 38:693.
Society of American Foresters. 1979. Forest Biomass as an Energy Source. Washington, D.C.
Springer, E. L., L. L. Zoch, Jr., W. C. Feist, G. J. Hajny, and R. W. Hemingway. 1978. Storage characteristics of southern pine whole tree chips. In Complete Utilization of Southern Pine, Charles W. McMillin, ed. Madison, Wis.: Forest Products Research Society.
Stake Technology, Ltd. 1979. Pro-Cell Nutritional Report No. 1. Ottawa, Ontario.
Stranks, D. W. 1959. Fermenting wood substrates with a rumen cellulolytic bacterium. For. Prod. J. 9:228.
Tarkow, H., and W. C. Feist. 1969. A mechanism for improving digestibility of lignocellulosic materials. In Cellulases and Their Applications. Advances in Chem. Ser. 95. Washington, D.C.: American Chemical Society.
Turner, H. D. 1964. Feed molasses from the Masonite Process. For. Prod. J. 14(7):282.
U.S. Department of Agriculture. 1960. Wood Molasses for Stock and Poultry Feed. Report No. 1731. Madison, Wis.: USDA Forest Products Laboratory.
U.S. Department of Agriculture. 1976. Utilization and marketing as tools for aspen management in the Rocky Mountains. Gen. Tech. Rep. RM-29. Fort Collins, Colo.: Rocky Mountain Forest and Range Experiment Station.
U.S. Department of Agriculture. 1977. The Nation's Renewable Resources—An Assessment, 1975. For. Resource Rep. 21. Washington, D.C.: Government Printing Office. 243 pp.
Van Soest, P. J. 1982. Limitations of ruminants. Pp. 336–339 in Nutritional Ecology of the Ruminant. Corvallis, Oreg.: O and B Books.
White, M. S. 1979. Guides for Reducing the Incidence of Fires During Outside Storage of Sawmill Residues. Blacksburg, Va.: Department of Forestry and Forest Products, Virginia Polytechnic Institute and State University.
Wilson, R. K., and W. J. Pigden. 1964. Effect of a sodium hydroxide treatment on the utilization of wheat straw and poplar wood by rumen microorganisms. Can. J. Anim. Sci. 44:122.
Woodward, F. E., H. T. Converse, W. R. Hale, and J. G. McNulty. 1924. Pp. 9–12 in Values of Feeds for Dairy Cows. USDA Bull. No. 1272. Washington, D.C.
Zerbe, J. I. 1977. Wood in the energy crisis. For. Farmer 37(2):12–15.

4
Animal Wastes

INTRODUCTION

Excreta contain several nutrients that are capable of being utilized when the material is recycled by feeding. Nitrogen, which is present in both protein and nonprotein forms, is a major constituent, and others are calcium and phosporus. Certain vitamins are synthesized in the intestine and appear in the excreta. The excreta also has energy value.

Coprophagy was recognized as a normal physiological phenomenon in rabbits (Madsen, 1939), and is natural in many wild and domestic species (Bjhornhog and Sjoblom, 1977). The first documented evidence of the importance of intestinal bacterial synthesis in nutrition was probably the work of Osborne and Mendel (1914), demonstrating that feeding 1 percent feces from normally fed rats to rats on a purified diet prevented death.

Bohstedt et al. (1943) found that cow manure had nutritional value for pigs, in addition to the grain it contained. Fuller (1956) reported that hydrolyzed poultry litter was as effective as fish meal in achieving growth from commercial type broiler diets.

Utilization of animal wastes as feedstuffs is not a new phenomenon. In 1925 Evvard and Henness reported that on the average one pig following 1.9 steers recovered the equivalent of 142 kg of corn during the 120-day feeding period. There have been a number of previous reviews on feeding animal waste (Anthony, 1971; Bhattacharya and Taylor, 1975; Blair and Knight, 1973; Fontenot and Jurubescu, 1980; Smith and Wheeler, 1979).

QUANTITY

Van Dyne and Gilbertson (1978) estimated that 112 million tons of manure dry matter were voided by farm animals in the United States in 1974, of which about 50 percent is recoverable. The quantities produced and collectable by classes of animals are shown in Table 29. Although cattle wastes account for about 80 percent of the total, only 41 percent is recoverable. Virtually all of the poultry waste is collectable. These estimates are lower than those of Heichel (1976), who estimated that 300 million tons of animal waste dry matter were produced per year, of which 50 percent was collectable. Other estimates were approximately 2 billion tons on a fresh basis (Taiganides and Stroshine, 1971; Wadleigh, 1968). Assuming a dry-matter content of 15 percent, this represents 300 million tons of dry matter.

The amount of animal-waste dry matter available annually in Canada is estimated to be 24 million tons (Pequegnat, 1975).

PHYSICAL CHARACTERISTICS

The wastes from farm livestock are of two types, solid and liquid. Generally, these are collected separately, unless automated systems involving pits are used. Poultry do not excrete urine separately, and the waste in this case is a semisolid mixture of feces and urine.

Because of the different systems of urinary excretion in livestock and poultry, a large part of the urinary nutrients tend to be lost from livestock wastes and retained with poultry wastes. Table 30 shows a partition of the nitrogen between feces and urine of various farm animals. An additional factor preventing loss of nitrogen in poultry waste is the form in which nitrogen is excreted in the urine, namely insoluble uric acid.

Wastes containing bedding or floor litter tend to be drier and may contain absorbed urinary waste. In the case of deep litter from poultry houses, the waste is frequently very dry and dusty. Handling of raw wastes tends to be difficult since they are either very wet or very dry.

In this report, cattle waste is defined as the solid waste from beef or dairy cattle. This waste may contain litter bedding material and appreciable levels of soil if animals are housed on dirt lots. Swine wastes from confinement herds are generally collected into pits and are frequently treated by a variety of aerobic systems to reduce odor and biological oxidation demand (BOD).

Poultry wastes are generally of two types, with or without litter material. The latter, from caged birds, is referred to in this report as caged layer waste. Commonly, it contains shed feathers, spilled feed, broken eggs, etc., in addition to excretory products. Litter waste is from birds grown

TABLE 29   Livestock and Poultry Waste Production in the United States, 1974

	Manure (1,000 tons, dry basis)	
Class of Animal	Production	Collectable
Beef cattle (range)	52,057	1,897
Feeder cattle	16,428	16,000
Dairy cattle	25,210	20,358
Hogs	13,360	5,538
Sheep	3,796	1,700
Laying hens	3,374	3,259
Turkeys	1,251	983
Broilers	2,086	2,434[a]
Total	111,562	52,169

[a]Includes litter.

SOURCE: Van Dyne and Gilbertson (1978).

under floor-raising systems and contains bedding material such as peanut or rice hulls, wood shavings, sawdust, or straw. Usually litter is obtained from broiler and turkey houses. Generally it contains some feathers and spilled feed. This type of waste is referred to in this report as poultry litter waste.

NUTRITIVE VALUE

Chemical Composition of Animal Wastes

Composition of animal wastes is shown in Appendix Tables 1 to 5. Most of the data refer to processed wastes, since raw wastes are generally unsuitable in recycling systems.

One notable feature of all of these wastes is variability in composition due to dietary regime, length of time before collecting, admixture with bedding, processing method, etc. Gilbertson et al. (1974) have outlined some of the factors affecting the nutrient and energy composition of beef cattle feedlot waste fractions.

Mean compositions of animal wastes (based on the data in Appendix Tables 1 to 5) are shown in Tables 31 to 34. They are characterized as having a relatively high content of crude protein (high nonprotein nitrogen) and a level of true protein that may be similar to that of the common feed grains. Other proximate constituents present at relatively high levels in

TABLE 30 Distribution of Nitrogen in Feces and Urine from Livestock

Species	Total Nitrogen (%)	
	Feces	Urine
Beef cattle	50	50
Dairy cattle	60	40
Sheep	50	50
Swine	33	67
Poultry	25	75

SOURCE: Smith (1973).

animal wastes are ash and fiber. Ether extract values are generally low. These features result in a relatively low level of available energy in animal wastes. The high ash content suggests that animal wastes are potentially good sources of minerals. Phosphorus is a valuable constituent.

These features indicate that animal wastes are more suited to recycling systems involving ruminants, since ruminants possess a digestive tract capable of efficiently utilizing fiber and nonprotein nitrogen. The wastes possessing the highest nutritive value appear to be broiler litter and layer waste.

The main difference in composition between raw and processed wastes is in moisture content; many of the processed wastes are low in moisture. Some volatile components, such as nitrogen, are also lower in some processed wastes because of losses during heating. In general, on a dry-matter basis, processed wastes share many of the characteristics of raw wastes.

Nutrient Utilization

Newton et al. (1977) conducted a nutritional evaluation of wastelage, a term introduced by Anthony (1970) to denote a mixture of fermented or ensiled cattle waste and a feedstuff such as forage containing 57 percent fresh cattle waste. Apparent digestibility of a control diet consisting of corn, Bermuda grass, and urea was 76.1 percent and did not differ significantly from that of a wastelage diet that was 73.7 percent. Dry-matter digestibility of fermented waste (by extrapolation) was 57.7 percent, nitrogen digestibility was 34.2 percent, crude fiber digestibility was 31.6 percent, and digestibility of nitrogen-free extract was 83.5 percent. Harpster et al. (1978) reported that nitrogen retention and digestibility of dry matter, organic matter, ether extract, nitrogen-free extract, and energy

TABLE 31 Mean Composition and Energy Value of Animal Wastes

		Composition of Dry Matter						Ruminant[a]				Swine[a]	Poultry[a]
Type of Resource	Dry Matter (%)	Crude Protein (%)	NPN (%)	Ash (%)	Ether Extract (%)	Fiber Crude (%)	Fiber ADF (%)	DDM (%)	TDN (%)	DE (Mcal/kg)		DE (Mcal/kg)	ME (Mcal/kg)
Cattle waste (processed)	20.4	16.9	—	14.0	3.14	23.9	40.3	42.5	—	—		—	—
Swine waste (processed)	—	27.9	0.86	18.0	7.0	15.0	26.5	—	—	—		2.13	—
Cage layer waste	30.8	33.9	3.65	22.4	2.12	19.1	17.2	78.7	—	—		—	—
Cage layer waste (processed)	85.7	30.0	2.75	30.0	2.18	13.1	—	56.6	—	—		—	1.149
Broiler waste (processed)	85.7	37.1	3.53	17.5	3.2	14.0	—	—	—	—		—	1.84
Broiler litter	83.0	29.7	2.16	16.7	3.3	19.8	—	67.4	—	—		—	—
Broiler litter (processed)	88.7	30.8	2.47	17.2	3.28	18.4	—	66.5	49.8	2.18		—	1.274

[a]Dry basis.

TABLE 32 Mean Mineral Composition of Animal Wastes (dry basis)

Type of Resource	Ash (%)	Ca (%)	P (%)	Na (%)	Cl (%)	Mg (%)	K (%)	S (%)	Fe (mg/kg)	Cu (mg/kg)	Co (mg/kg)	Mn (mg/kg)	I (mg/kg)	Zn (mg/kg)	Se (mg/kg)	Mo (mg/kg)
Cattle waste (processed)	18.9	2.34	1.03	0.88	1.32	0.4	0.5	—	1340.6	31.0	—	147.5	—	242.4	—	—
Swine waste (processed)	18.0	3.04	2.59	2.75	—	1.1	1.9	0.3	3723.5	114.2	6.1	342.0	—	709.4	—	0.3
Cage layer waste	22.0	7.08	1.74	0.28	—	0.51	2.05	—	662.1	31.1	—	201.2	—	208.9	—	—
Cage layer waste (processed)	30.4	8.13	2.22	0.46	1.01	0.65	1.63	—	1773.5	70.3	1.9	373.7	—	477.2	0.56	—
Broiler waste (processed)	17.5	3.4	2.0	0.47	—	0.53	1.38	—	1690	32	—	432	—	326	—	—
Broiler litter	16.9	2.3	1.68	—	—	—	—	—	—	—	—	—	—	—	—	—
Broiler litter (processed)	15.0	2.29	1.81	0.38	—	0.46	2.16	—	—	23	—	—	—	343	—	—

TABLE 33 Mean Additional Mineral Composition of Animal Wastes

Type of Resource	Sm (%)	Th (%)	U (%)	Yb (%)	Sb (%)	As (%)	Br (%)	Fl (%)	Al (%)	Cd (%)	Pb (%)
Swine waste (processed)	—	—	—	—	—	—	—	—	54.4	1.0	12.1
Cage layer waste (processed)	—	—	—	—	—	2.2	18.7	35	—	.85	3.0
Broiler litter	—	—	—	—	—	6.2	—	—	—	—	—
Broiler litter (processed)	—	—	—	—	—	19.4	—	—	—	—	—

were lower in steers when the diet consisted entirely of ensiled cattle waste. Addition of high-moisture corn increased the means for these parameters.

Lucas et al. (1975) reported that dehydrated cattle waste had a dry-matter digestibility of 16.6 percent and that the metabolizable energy value was 0.485 Mcal/kg dry matter.

Thorlacius (1976) reported a nutritional evaluation of dehydrated cattle waste using sheep. Diets containing 0, 50, or 100 percent waste were compared. During the final 10 days of a digestibility trial, intake of dry matter was 2,632, 2,277, and 2,050 g for the three diets, respectively; dry-matter digestibility was 62.7, 51.7, and 26.7 percent, respectively; and nitrogen digestibility of the three diets was 70.9, 62.6, and 42.2 percent, respectively.

Lipstein and Bornstein (1971) investigated the value of dried cattle waste for broiler chickens. It was concluded that the waste had little or no value as a source of energy or protein. Littlefield et al. (1973) found that cattle waste was of some value as a dietary source of yolk pigment for laying hens. Faruga et al. (1974) also reported the yolk color was improved with cattle waste in the diet of laying hens. With 0, 3, 6, and 9 percent dehydrated waste in the diet, feed intake per dozen eggs was 5.83, 5.49, 6.03, and 5.20 kg or 0.99, 0.94, 1.04, and 0.88 kg digestible crude protein/kg eggs, respectively. Metabolizable energy was 14.16, 13.66, 15.36, and 13.59 Mcal/kg eggs, respectively. The nutritional value for poultry of a protein fraction of cattle waste obtained by a commercial process (Cereco) was reported by Kienholz et al. (1975). Estimated metabolizable energy value (poultry) of the product was 2.3 Mcal/kg.

Hennig et al. (1972) conducted a nutritional evaluation of dehydrated swine waste using cattle and sheep. With sheep, digestibility of waste nitrogen in a pelleted diet was 57.4 percent. Cattle fed pellets containing 40 percent swine waste plus 1 kg hay daily ate on average 6.13 kg pellets

TABLE 34 Mean Amino Acid Composition of Animal Wastes (dry basis)

Type of Resource	Crude Protein	Arg (%)	His (%)	Iso (%)	Leu (%)	Lys (%)	Met (%)	Cys (%)	M+C (%)	Phe (%)	Tyr (%)	P+T (%)	Thr (%)	Try (%)	Val (%)
Cattle waste (processed)	19.4	0.17	0.11	0.22	0.56	0.42	0.08	—	—	0.04	0.13	—	0.26	—	0.35
Swine waste (processed)	26.6	0.7	0.28	0.78	1.41	0.92	0.44	0.33	—	0.67	0.65	—	0.9	0.28	0.97
Cage layer waste (processed)	31.8	0.61	0.28	0.5	0.86	0.57	0.15	0.16	0.39	0.52	0.35	0.9	0.58	0.34	0.64
Broiler waste (processed)	43.3	0.81	0.29	0.57	1.04	0.71	0.27	0.4	—	0.63	0.52	—	0.73	—	0.77
Broiler litter (processed)	33.9	0.59	0.26	0.65	1.03	0.64	0.16	0.07	0.87	0.55	0.35	1.16	0.61	0.39	0.85

daily and gained 1.1 kg daily. Flachowsky (1975) fed mixtures of pelleted feed containing 30 or 50 percent swine waste to growing bulls. After a period of adjustment the diets were accepted readily and consumption of the test diets was 7.58 and 8.20 kg dry matter/head/day, respectively. Solids obtained from semiliquid swine waste were found to have a lower energy value than the waste itself, and the feeding value was estimated to be similar to that of medium quality hay. Flachowsky (1977) also reported on another experiment on the nutritional value of the undissolved fraction of swine waste. Intake was depressed when the waste represented 50 percent of the diet, which was attributed to the high iron content of the diet (2,002 mg/kg). Pearce (1975) found that the dry-matter digestibility of swine waste was about 29 percent when fed to steers or sheep at levels up to 45 percent of the diet as a replacement for hay. Kornegay et al. (1977) reported that the digestibility of the components of swine waste when fed to growing gilts were energy, 46.7 percent; dry matter, 48 percent; crude protein, 60.1 percent; crude fiber, 40.9 percent; ether extract, 54.1 percent; nitrogen-free extract, 45.9 percent; and ash, 31.6 percent. Jentsch et al. (1977) investigated the energy value of pelleted swine waste with young bulls and mature sheep. Digestibility was only slightly reduced when the level of waste in the diet was raised from 25 to 42 percent. Addition of waste to the diet increased the proportion of butyric acid in rumen fluid in both species.

Oltjen and Dinius (1976) investigated the nutritive value for cattle of cage layer waste dehydrated and processed to recover compounds for industrial and medical use. Diets did not differ in digestibility of dry matter or acid detergent fiber when they contained 0 or 10.5 percent waste. Nitrogen retention was 30.1 percent with the waste diet, 34.1 percent with 3 percent uric acid in the diet, and 41.4 percent with 3.5 percent sodium urate in the diet. Rumen fluid pH was between 6.7 and 7.1 with all diets and was not affected by treatment. With up to 15 percent waste in the diet there was no significant difference in digestibility of dry matter, fiber, energy, or nitrogen or in nitrogen retention.

With dairy cows, Kristensen et al. (1976) reported that the digestibility of organic matter in dehydrated layer waste was 60 to 65 percent. Silva et al. (1976) conducted a digestibility trial with dairy cows fed diets containing up to 30 percent dehydrated layer waste. Digestibility of energy declined from 59.5 percent with the control to 54.8 percent with the diet containing 30 percent waste, and crude protein digestibility dropped from 59.2 to 53.1 percent. Digestibility of dry matter with the two diets was 58.5 and 39 percent, respectively. Including 10 percent waste in the diet had a slight effect on digestibility of dry matter and crude protein, and there was no marked reduction until the level of inclusion reached 20 percent.

Guedas (1966, 1967) reported on the digestibilities of diets containing 0 or 80 percent dehydrated layer waste with sheep. Mean digestibility of the waste diet was crude protein, 69.9 percent; crude fiber, 29.9 percent; and nitrogen-free extract, 71.4 percent. It was calculated that uric acid represented about 8 percent of the absorbed nitrogen. Parigi-Bini (1969) reported that the metabolizable energy value of dehydrated layer waste for sheep was 2.22 Mcal/kg. With 0 or 32 percent waste in the diet, apparent digestibilities were, respectively, dry matter, 87.5 and 80.5 percent; crude protein, 85.0 and 77.9 percent; crude fiber, 59.4 and 46.7 percent; ether extract, 85.7 and 78.7 percent; and nitrogen-free extract, 93.1 and 89.1 percent. Neither rumen pH nor the molar proportions of volatile fatty acids in rumen liquor were affected. Diets containing 0, 25, 50, 75, or 100 percent dehydrated layer waste were fed to sheep (Lowman and Knight, 1970). Metabolizable energy value of the waste was estimated to be 1.74 Mcal/kg dry matter. Values calculated for the diets containing 100 percent waste and 100 percent barley were, respectively, dry matter, 56.6 and 77.9 percent; organic matter, 66.5 and 80.7 percent; energy, 60.3 and 80.0 percent; nitrogen, 77.2 and 68.4 percent; and copper, 24.2 and 51.0 percent.

Tinnimit et al. (1972) fed diets in which dehydrated layer waste or soybean meal supplied 40 to 65 percent of total protein to sheep averaging 31 kg. When waste was increased in the diet from a level of 20 to 80 percent, dry-matter digestibility fell from 74 to 58 percent, and organic-matter digestibility fell from 77 to 68 percent. It was calculated that the digestible dry matter in dehydrated layer waste was about the same as that in low-quality alfalfa hay but that digestible organic matter was about 1.25 times greater. Bohme (1973) reported that with sheep the digestibility of organic matter in dehydrated layer waste was 67 percent. The digestible energy value of the waste was estimated to be 2.304 Mcal/kg dry matter and TDN value was estimated at 51.4 percent. With 30 percent dehydrated layer waste in the diet, Hennig et al. (1975) reported that crude protein digestibility of the waste by sheep was 83 percent and that digestible crude protein content was 33 percent (dry-matter basis). Digestibility of dehydrated layer waste for sheep was estimated by Salo et al. (1975) as follows: organic matter, 62.8 percent; crude protein, 76.9 percent; ether extract, 33.1 percent; crude fiber, 31.7 percent; and nitrogen-free extract, 57.5 percent. Metabolizable energy value was estimated to be 1.58 Mcal/kg. High values for digestibility of organic matter in sheep fed dehydrated layer waste were reported by Zgajnar (1975). The diets contained 0, 15, 25, or 35 percent waste. The effect of raising the level of waste in the diet was to lower the organic-matter digestibility of waste from 97.6 to 84 percent. Digestibility for crude protein increased from 68.6 to 77.9 percent and for nitrogen-free extract from 79.9 to 87.6 percent as the level of waste was increased from 15 to 35 percent. Digestibility of crude fiber was 85.5 percent when the diet contained 25 percent waste, and it was lower with the other levels. Smith and Lindahl (1977) found that lambs digested the dietary nutrients equally well when the diet contained alfalfa or dehydrated layer waste, except that ash was 43 percent less digestible in waste diets. Swingle et al. (1977) found that the dry-matter intake of sheep was not affected when dehydrated poultry waste, cottonseed meal, or urea provided over 85 percent of total dietary nitrogen. About 35 percent of the absorbed nitrogen was retained with the diet containing cottonseed meal, whereas 16 percent was retained with the diet containing dehydrated layer waste or urea.

Using dehydrated layer waste ensiled with corn forage, Goering and Smith (1977) found that digestibility of dry matter by sheep was 63 percent; it was 64 percent with a control diet containing soybean meal. Organic-matter digestibility was the same at 65 percent. Daily organic matter consumed was 35 and 28 g/Wkg$^{0.75}$, respectively.

Smith and Calvert (1976) compared dehydrated broiler waste and soybean meal as nitrogen supplements for sheep. Digestibilities of dry matter, organic matter, and nitrogen were not significantly different at 65.4 and 65.2 percent, 66.4 and 65.4 percent, and 53.7 and 57.9 percent, respectively.

Poultry litter waste was evaluated as a nutrient by several investigators. Bhattacharya and Fontenot (1965) found with sheep that the digestibility of crude protein in broiler litter was 64.8 to 67.1 percent. Apparent digestibilities of peanut hull and wood shaving broiler litter with sheep were: crude protein, 70.4 to 73.5 percent; crude fiber, 66.1 to 71.5 percent; ether extract, 56.3 to 62.7 percent; nitrogen-free extract, 68.6 to 74 percent; dry matter, 61.5 to 66.1 percent; and energy, 63.1 to 64.8 percent (Bhattacharya and Fontenot, 1966). Geri et al. (1970a,b,c) reported that the production of volatile fatty acids in vitro from litter waste was similar to that from common feedstuffs, with a tendency towards an increased proportion of propionic acid. Protein digestibility of the waste was estimated to be at least 87 percent.

Cross et al. (1978) found with beef steers that blood plasma urea nitrogen was about 50 percent higher with a diet containing 50 percent broiler litter silage than with a control diet, but that plasma and rumen fluid values were within normal physiological ranges. Muftic et al. (1968) fed a diet containing 80 percent poultry litter waste to dairy cows and reported that digestibilities were: dry matter, 60.3 percent; crude protein, 63.8 percent; true protein, 62 percent; nonprotein nitrogen, 71.4 percent; fiber, 26.6 percent; and nitrogen-free extract, 69.9 percent. Digestible crude protein of the litter was estimated to be 16.2 percent and starch equivalent was 35.8 percent.

Unprocessed wastes do not appear to have been evaluated as animal feeds, but investigations have been conducted on wastes with minimal processing. Flipot et al. (1975) reported on an evaluation of fresh poultry waste treated with 3 percent tannic acid or 2 percent paraformaldehyde and included in the diet of growing sheep at a level of 64 percent, wet-weight basis. Digestibility of dry matter and total nitrogen was 69.9 and 80.8 percent for a control diet based on soybean meal, 59.3 and 71.8 percent for the diet with tannic acid, and 54.8 and 71.5 percent for the diet with paraformaldehyde, respectively. Except for total nitrogen digestibility these differences were significant with the waste diets. Dry-matter intake was lower with the paraformaldehyde diet.

Evans et al. (1978b) fed wet cage layer waste along with corn silage to yearling Suffolk ewes. The waste was offered unsupplemented or supplemented with 2 percent molasses or 1 percent propionate. The effect of supplementation was to increase intake of waste dry matter from about 18.5 to 19.5 g/Wkg$^{0.75}$/day with molasses and to about 23 g/Wkg$^{0.75}$/day with propionate. On all treatments, intakes of total nitrogen and minerals were in excess of accepted requirements, indicating that the waste was a good source of these nutrients.

Smith et al. (1978) investigated wet cage layer waste as a protein supplement for calves. It was concluded that plasma urea nitrogen levels indicated a relatively slow release of nitrogen from the waste, which the researchers considered made it an ideal source of nitrogen for rumen function.

Performance of Animals Fed Animal Wastes

Cattle Waste

Anthony and Nix (1962) and Anthony (1966) established the feasibility of feeding steer waste to cattle. Anthony (1970) fed yearling beef animals a concentrate diet consisting of corn silage and ground ear corn supplemented with urea, cottonseed meal, minerals, and vitamin A, or a diet consisting of 40 parts wet cattle waste and 60 parts air-dried concentrate. No effect on feed intake or gain was noted. Hsu (1976) reported on the performance of cattle feed silages containing 0 to 36 percent cattle waste. In one trial, daily gain was decreased by 50 percent as a result of the inclusion of waste in the diet. In the other trial, daily gain was 1.22 kg with the waste and 1.14 with the control diet.

Newton et al. (1977) fed heifers a control diet of corn, Bermuda grass pellets, and urea, or a wastelage diet containing 40 percent wet fermented cattle waste. Control animals averaged 1.34 kg/day and required 5.02 kg dry matter/kg gain; wastelage animals averaged 1.27 kg/day and required

5.40 kg dry matter/kg gain. Harpster et al. (1978) investigated the growth of steers fed diets containing 40 to 75 percent ensiled cattle waste, the remainder of the ration being high-moisture corn. Daily gain and feed/gain ratio for the control diet were 1.10 kg and 6.48. With 40, 50, and 75 percent waste in the diet these parameters were 1.05 kg and 7.88; 1.03 kg and 7.95; and 0.75 kg and 10.41, respectively. Lamm et al. (1979) found that dry-matter intake was similar for calves fed wastelage or wastelage treated with sodium hydroxide, but was lower than that of calves fed a corn silage control diet. Daily gain was similar with all diets. Richter et al. (1980) investigated performance of cattle fed 0, 20, 40, or 60 percent cattle waste roughage obtained by a commercial process (Corral Industries Inc.). Over a 124-day period daily gains were 1.10, 1.53, 1.58, and 1.51 kg, respectively. Schake et al. (1977) found that body weight maintenance was achieved in cows when 68.6 percent of the feed dry matter was high-fiber waste obtained from fresh cattle waste by use of a vibrating screen technique.

Diaz and Elias (1976) reported on the use of cattle waste in swine feeding. Pigs averaging 30 kg were fed diets supplemented with torula yeast and containing 0, 25, or 50 percent ensiled cattle waste. Daily gain was 507, 419, and 375 g, respectively, and feed/gain ratio was 3.99, 4.65, and 5.57.

Laying hens were fed a control diet or a diet containing 10 percent dehydrated cattle waste as the only source of animal protein (Saedi and Zohari, 1968). Egg production was 70 and 61.7 percent, respectively, and feed/dozen eggs was 3.70 and 4.12 kg.

Swine Waste

Flachowsky (1975, 1977) investigated the performance of cattle fed diets containing swine waste. Mixtures of pelleted feed containing 30 or 50 percent solid material from swine semiliquid waste were fed to cattle in a 252-day test. A control group was given pelleted feed with 36 percent straw. Daily gain for the control and groups fed 30 and 50 percent waste was 1.23, 1.18, and 1.00 kg, respectively. In another trial, bulls were grown for 315 days on pelleted diets containing 0, 25, or 50 percent swine slurry waste as a replacement for straw. Daily gain was 1.03, 1.04, and 0.84 kg, respectively, and intake of dry matter was 4.60, 3.23, and 3.47 kg/kg gain.

Poultry Waste

Performance data for cattle fed diets containing dehydrated layer waste (DLW) are summarized in Table 35 (Andersen et al., 1976; Batsman,

TABLE 35 Performance of Cattle Fed Diets Containing Dehydrated Layer Waste (DLW)

	Dietary Treatment		
Performance	Control	Waste	Reference
Daily gain (kg)	1.22	1.22	El-Sabban et al. (1970)
	1.24	1.25	Meregalli et al. (1971)
	1.42	1.47	Meregalli et al. (1973)
	0.98	1.01	Pereira et al. (1972)
	0.79	0.70	Tinnimit et al. (1972)
	0.79	0.81	Batsman (1973)
	1.20	1.22	Oliphant (1974)
	0.81	0.91	Clark et al. (1975)
	1.43	1.31	Andersen et al. (1976)
	1.20	1.11	Cullison et al. (1976)
	0.72	0.68	Lamm et al. (1976)
	0.60	0.86	Oltjen and Dinius (1976)
	1.62	1.29	Koenig et al. (1978)
	0.98	0.93	Smith et al. (1979)
	1.01	0.95	Vijchulata et al. (1980)
Mean	1.07	1.05	
Daily feed (kg DM)	10.27	10.02	El-Sabban et al. (1970)
	2.56	2.50	Tinnimit et al. (1972)
	5.77	5.75	Oliphant (1974)
	7.80	8.70	Clark et al. (1975)
	8.74	8.77	Cullison et al. (1976)
	5.70	5.47	Lamm et al. (1976)
	9.35	7.43	Koenig et al. (1978)
	6.57	6.51	L. W. Smith et al. (1979)
	7.95	7.99	Vijchulata et al. (1980)
Mean	7.19	7.02	
Feed/gain ratio	12.53	12.12	El-Sabban et al. (1970)
	5.96	6.40	Meregalli et al. (1971)
	5.78	5.61	Meregalli et al. (1973)
	13.20	13.80	Pereira et al. (1972)
	3.24	3.50	Tinnimit et al. (1972)
	4.79	4.88	Oliphant (1974)
	9.60	9.60	Clark et al. (1975)
	3.77	3.78	Andersen et al. (1976)
	7.28	7.90	Cullison et al. (1976)
	7.94	8.00	Lamm et al. (1976)
	14.18	10.32	Oltjen and Dinius (1976)
	6.07	5.91	Koenig et al. (1978)
	7.16	7.81	L. W. Smith et al. (1979)
	7.85	8.45	Vijchulata et al. (1980)
Mean	7.81	7.72	

1973; Clark et al., 1975; Cullison et al., 1976; El-Sabban et al., 1970; Koenig et al., 1978; Lamm et al., 1976; Meregalli et al., 1971, 1973; Oliphant, 1974; Oltjen and Dinius, 1976; Pereira et al., 1972; Tinnimit et al., 1972; and Vijchulata et al., 1980). The means for waste-fed animals were obtained by averaging over all levels, though some of the levels (up to 50 percent of the dry matter) were probably excessive. Mean daily liveweight gain, daily feed dry-matter intake, and feed/gain ratio for animals fed control and waste diets were, respectively, 1.07 kg and 1.05, 7.19 kg and 7.02, and 7.81 kg and 7.72. These results indicate excellent performance with diets containing waste.

The data obtained by Bucholtz et al. (1971) were not included with the above data because of the unusually low crude protein content (17 percent) of the waste used and the anticipated poor performance obtained with a level of 32 percent waste in the diet. In addition these workers reported that the test animals selectively avoided the waste in the diet, with the result that the dry matter consumed had a low crude protein content. Generally, the inclusion of poultry waste did not affect feed intake adversely. However, Koenig et al. (1978) reported a reduction in intake from 9.35 to 7.43 kg dry matter daily when 10 percent of a formaldehyde-treated waste was fed. This was attributed to the waste diet's becoming stale quickly at the high environmental temperatures experienced. Such temperatures could also be expected to volatilize any residual formaldehyde, with an adverse effect on intake.

A waste used by Andersen et al. (1976) was Urimix, a commercial product available in Scandinavia, containing 90 percent dehydrated layer waste, 5 percent animal fat, and 5 percent molasses. Up to 45 percent was included in the diet of young bulls with no loss in performance until the content of Urimix in the diet exceeded 30 percent. The remainder of the diet was barley, oats, vitamins, and minerals, and in some cases soybean meal.

Cullison et al. (1976) reported on the use of dehydrated broiler waste (DBW) in steer diets. Three diets based on corn were employed: control (soybean meal supplying the supplementary protein), half of the supplementary protein supplied by dehydrated broiler waste, and all of the supplementary protein supplied by DBW. Respective daily gains were 1.20, 1.18, and 1.11 kg; daily feed intake was 8.74, 8.89, and 8.77 kg; and feed/gain ratio was 7.28, 7.53, and 7.90.

Milk-production data are summarized in Table 36 for dairy cows fed DLW at levels generally about 15 percent of the dietary dry matter (Bull and Reid, 1971; Kneale and Garstang, 1975; Kristensen et al., 1976; Silva et al., 1976; Smith et al., 1976; Thomas et al., 1972). The means for waste-fed cows were obtained by averaging overall levels, though some

TABLE 36 Milk Production of Cows Fed Diets Containing Dehydrated Layer Waste (DLW)

Performance	Dietary Treatment Control	Waste	Reference
Milk yield (kg/day)	21.2	17.8	Bull and Reid (1971)
	19.6	20.6	Thomas et al. (1972)
	14.8	15.2	Kneale and Garstang (1975)
	20.0	19.8	Kristensen et al. (1976)
	21.2	17.2	Silva et al. (1976)
	17.1	15.4	Smith et al. (1976)
Mean	19.0	17.7	
Milk fat (%)	3.68	3.92	Bull and Reid (1971)
	3.30	3.87	Thomas et al. (1972)
	3.47	3.42	Kneale and Garstang (1975)
	3.41	3.33	Silva et al. (1976)
	3.70	3.60	Smith et al. (1976)
Mean	3.51	3.63	
Milk total solids (%)	12.40	12.56	Bull and Reid (1971)
	11.80	11.85	Kneale and Garstang (1975)
	11.93	11.62	Silva et al. (1976)
Mean	12.04	12.01	

of the levels of waste fed were probably excessive. Mean daily milk yield with fat and total solids for cows fed control and waste diets were, respectively, 19 and 17.7 kg, 3.51 and 3.63 percent, and 12.04 and 12.01 percent. The overall trend was for daily milk yield to be reduced slightly, but the results are very encouraging since it is well known that milk production of dairy cows can easily be depressed by the use of inappropriate diets. Mean fat content was slightly higher in waste-fed cows, possibly in response to the reduction in total milk yield.

A waste product used by Kristensen et al. (1976) was Urimix similar to that used by Andersen et al. (1976). The highest level used was 40 percent of the concentrate portion of the diet, and in general, the differences in milk production between cows fed that diet and a control diet were not significant. The beneficial effect on energy intake of including animal fat in the Urimix product probably explains why milk production was not affected in the experiments conducted by Kristensen et al. (1976), yet was lowered with the dehydrated layer waste at up to 32 percent in dairy concentrates in the experiments conducted by Kneale and Garstang (1975) and Smith et al. (1976).

Performance data for sheep fed diets containing dehydrated layer waste are summarized in Table 37 (Cuevas, 1969; Goering and Smith, 1977; Kazheka and Kozyr, 1975; Merwe et al., 1975; Smith and Calvert, 1976; Smith and Lindahl, 1977; Thomas et al., 1972; Tinnimit et al., 1972). The means for waste-fed animals were obtained by averaging overall levels, though some of the levels (up to 50 percent of the dry matter) were probably excessive. Mean daily liveweight gain and feed/gain ratio for animals fed control and waste diets were, respectively, 0.19 and 0.18 kg, and 5.52 and 6.66. These results indicate excellent growth performance in growing sheep fed diets containing waste, though utilization of diets containing waste appears to be significantly lower than that of control diets. This may be a reflection of inadequate energy and/or excess ash for young growing lambs.

TABLE 37 Performance of Sheep Fed Diets Containing Dehydrated Layer Waste (DLW)

Performance	Control	Waste	Reference
Daily gain (kg)	0.20	0.22	Cuevas (1969)
	0.21	0.16	Thomas et al. (1972)
	0.35	0.24	Tinnimit et al. (1972)
	0.22	0.21	Berbeci et al. (1975)
	0.18	0.17	Kazheka and Kozyr (1975)
	0.20	0.12	Merwe et al. (1975)
	0.05	0.13	Goering and Smith (1977)
	0.19	0.18	Smith and Calvert (1976)
	0.15	0.18	Smith and Lindahl (1977)
Mean	0.19	0.18	
Feed/gain ratio	7.14	11.11	Thomas et al. (1972)
	4.16	4.52	Tinnimit et al. (1972)
	4.50	3.41	Berbeci et al. (1975)
	6.53	10.30	Merwe et al. (1975)
	5.87	6.53	Smith and Calvert (1976)
	4.94	4.08	Smith and Lindahl (1977)
Mean	5.52	6.66	

Pelleting of the feed appears to be beneficial in avoiding low intake. Smith and Calvert (1976) and Smith and Lindahl (1977) used complete pelleted diets and found that differences in performance criteria were not significant.

Wet poultry wastes have been used in several trials. McNiven et al. (1976) mixed unprocessed layer waste containing 75 percent moisture in a diet diluted to 80 percent moisture for sheep. A control diet had either 10 or 80 percent moisture. With the two control and the waste diets, respectively, daily gain was 0.12, 0.14, and 0.13 kg, and intake of organic matter/kg gain was 6.93, 5.34, and 6.03 kg. Smith et al. (1978) included 0 or 22 percent fresh waste from caged layers in the diet of bull calves. The diet was based on corn silage and high-moisture corn, and was supplemented with soybean meal or waste. Daily gain was 1.29 and 1.03 kg, respectively, and daily dry-matter intake was 4.15 and 3.86 kg.

In a subsequent investigation fresh cage layer waste was treated with a 0.5 percent wet weight acetic-propionic acid mixture, stockpiled and mixed at a 68 percent level (wet basis) into corn silage just prior to feeding. The diet was fed to growing/finishing Hereford steers, in comparison with diets based on soybean meal or urea. Daily gain, feed efficiency, and carcass and empty body gains were similar with all treatments. It was concluded that up to 22 percent waste (DM basis) was readily accepted by cattle.

Only a few experiments on feeding dehydrated layer waste have involved swine, and performance data are summarized in Table 38 (Denisov et al., 1974, 1975b; Geri, 1968; Osterc, 1972). The means for waste-fed

TABLE 38 Performance of Swine Fed Diets Containing Dehydrated Layer Waste (DLW)

Performance	Control	Waste	Reference
Daily gain (kg)	0.55	0.49	Geri (1968)
	0.65	0.51	Osterc (1972)
	0.61	0.52	Denisov et al. (1974)
	0.59	0.55	Denisov et al. (1975b)
Mean	0.60	0.52	
Feed/gain ratio	3.66	3.99	Geri (1968)
	4.58	5.64	Denisov et al. (1974)
Mean	4.12	4.82	

animals were obtained by averaging overall levels. Mean daily liveweight gain and feed/gain ratio for animals fed control and waste diets were, respectively, 0.60 and 0.52 kg, and 4.12 and 4.82 kg. These results are encouraging since levels of up to 35 percent waste were used and since the pig does not have a digestive system capable of utilizing crude fiber or nonprotein nitrogen efficiently.

One interesting aspect of the experiments conducted by Geri (1968) is that the swine given dehydrated layer waste received no supplements of antibiotic or vitamin B_{12}. Also, the initial weights of the pigs involved were as low as 17 kg in some groups.

Performance data for growing chickens fed diets containing dehydrated layer waste are summarized in Table 39 (Biely and Stapleton, 1976; Biely et al., 1972; Flegal and Zindel, 1971; Fookes, 1972; Lee and Blair, 1973; Lee and Yang, 1975; Stapleton and Biely, 1975). The means for waste-fed chickens were obtained by averaging overall levels, although some of the levels (up to 30 percent) were excessive. Mean daily liveweight gain and feed/gain ratio for chickens fed control and waste diets were, respectively, 16.1 and 15.7 g, and 2.36 and 2.60 g. These results indicate that growth can be maintained with diets containing waste, but that feed efficiency is depressed. This can be attributed mainly to the low metabolizable energy value of the waste. Flegal and Zindel (1971) reported that growth and feed/growth ratio with diets containing waste could be improved by the addition of fat to the diet.

Bhargava and O'Neil (1975) investigated the growth performance of broiler chickens fed diets containing up to 20 percent dried waste from caged broilers. Mean daily liveweight gain and feed/gain ratio for the control and waste-fed broilers were, respectively, 23.6 and 23.0 g, and 2.28 and 2.33 g. All levels above 5 percent waste reduced gain and feed efficiency significantly, but levels up to 20 percent had no significant effect on growth when the diets were equalized for energy and protein. The data of Trakulchang and Balloun (1975a,b) were not included in the above summary since the waste used was an air-dried product with a very low level of crude protein (21 percent). They found that the inclusion of broiler waste in the diet of growing chickens at a level of 10 or 20 percent resulted in a progressive decrease in growth rate and feed-conversion efficiency.

Performance data for laying chickens fed diets containing dehydrated layer waste are summarized in Table 40 (Flegal and Zindel, 1971; Hodgetts, 1971; Lee and Bolton, 1977; Lee and Yang, 1976; Vogt, 1973). The means for waste-fed birds were obtained by averaging overall levels, though some of the levels used (up to 40 percent) were excessive. Mean

TABLE 39 Performance of Growing Chickens Fed Diets Containing Dehydrated Layer Waste (DLW)

	Dietary Treatment		
Performance	Control	Waste	Reference
Daily gain (g)	15.6	15.4	Flegal and Zindel (1971)
	16.9	15.8	Biely et al. (1972)
	7.7	7.7	Fookes (1972)
	29.3	30.3	Lee and Blair (1973)
	24.6	24.5	Lee and Yang (1975)
	10.1	8.8	Stapleton and Biely (1975)
	8.7	7.7	Biely and Stapleton (1976)
Mean	16.1	15.7	
Feed/gain ratio	2.11	2.25	Flegal and Zindel (1971)
	2.43	2.81	Biely et al. (1972)
	2.87	3.23	Fookes (1972)
	2.50	2.42	Lee and Blair (1973)
	2.61	2.92	Lee and Yang (1975)
	2.05	2.29	Stapleton and Biely (1975)
	1.98	2.28	Biely and Stapleton (1976)
Mean	2.36	2.60	

egg production and feed/dozen eggs for the control and waste-fed birds were, respectively, 71.9 percent and 72.8 kg, and 1.90 percent and 1.90 kg.

Flegal and Zindel (1971) reported that the inclusion of up to 20 percent waste did not influence egg production or feed efficiency, and that the inclusion of up to 40 percent did not adversely affect egg weight or shell thickness. Vogt (1973) concluded that 10 percent waste was unsuitable for inclusion in layer feed unless the energy content of the feed was raised.

Lee et al. (1976) raised replacement layers on diets containing up to 5 percent dehydrated layer waste. No effects on weight at 18 weeks were noted, but feed consumption and feed efficiency were better with the control diet. Subsequent laying performance was not affected.

Poultry Litter

Performance data for cattle fed diets containing poultry litter are summarized in Table 41 (Batsman, 1973; Borgioli and Tocchini, 1969; Bosman, 1973; Boubedja and Marx, 1974; Cross and Jenny, 1976; Cross et al., 1978; Cullison et al., 1976; Denisov et al., 1973; Fontenot et al.,

TABLE 40 Performance of Laying Hens Fed Diets Containing Dehydrated Layer Waste (DLW)

	Dietary Treatment		
Performance	Control	Waste	Reference
Egg production	64.7	61.1	Flegal and Zindel (1971)
	78.9	79.6	Hodgetts (1971)
	82.3	86.9	Vogt (1973)
	73.8	72.4	Lee and Yang (1976)
	59.6	64.0	Lee and Bolton (1977)
Mean	71.9	72.8	
Kg feed/dozen eggs	1.95	2.17	Flegal and Zindel (1971)
	2.00	1.86	Hodgetts (1971)
	1.63	1.51	Vogt (1973)
	1.64	1.66	Lee and Yang (1976)
	2.29	2.30	Lee and Bolton (1977)
Mean	1.90	1.90	

1966, 1971a; Kanev et al., 1971; Meregalli et al., 1973; Noland et al., 1955; Sommer and Pelech, 1971; Szelenyi et al., 1971; Velloso et al., 1970). The means for litter-fed animals were obtained by averaging over-all levels, though some of the levels used (up to 60 percent) were probably excessive. Mean daily gain and feed/gain ratio for control and litter-fed animals were 1.00 and 0.87 kg, and 10.18 and 11.58 kg, respectively.

No significantly depressing effects on bull performance as a result of including up to 50 percent poultry litter were reported by Kanev et al. (1971). Daily gain and feed/gain ratio with the control and test diets were, respectively, 1.12 kg and 5.49 feed units/kg gain, and 1.11 kg and 6.76 feed units/kg gain. Szelenyi et al. (1971) reported that cattle fed for 3 months on a standard grain mixture gained 1.25 kg daily, but that daily gain fell to 0.93 kg when 25 percent of the mixture was replaced by poultry litter. In a second trial the cattle gained 1.22 kg daily with a mixture containing 25 percent litter and adjusted levels of bran and peanut meal. Levels of 33 and 50 percent of the diet (dry-matter basis) were used successfully with Simmental bulls (Batsman, 1973). Daily gain with a control and with the two test diets was 0.87, and 0.90 and 0.87 kg, respectively. In a subsequent trial, bulls 7 months of age were fed a control diet or diets containing 20, 40, or 60 percent litter. Daily gain was 0.71, 0.75, 0.71, and 0.72 kg, respectively. No significant effects on carcass

TABLE 41 Performance of Cattle Fed Diets Containing Poultry Litter

Performance	Dietary Treatment Control	Waste	Reference
Daily gain (kg)	0.92	0.77	Noland et al. (1955)
	0.73	0.52	Fontenot et al. (1966)
	1.32	1.24	Borgioli and Tocchini (1969)
	0.90	0.77	Velloso et al. (1970)
	0.73	0.52	Fontenot et al. (1971a)
	1.12	1.11	Kanev et al. (1971)
	1.14	0.94	Sommer and Pelech (1971)
	1.25	1.07	Szelenyi et al. (1971)
	0.79	0.79	Batsman (1973)
	1.61	1.01	Bosman (1973)
	0.90	0.81	Denisov et al. (1973)
	1.32	1.23	Meregalli et al. (1973)
	0.96	0.70	Boubedja and Marx (1974)
	0.42	0.51	Cross and Jenny (1976)
	1.17	1.12	Cullison et al. (1976)
	0.72	0.83	Cross et al. (1978)
Mean	1.00	0.87	
Feed/gain ratio	13.16	17.38	Noland et al. (1955)
	13.10	15.60	Fontenot et al. (1966)
	13.10	16.40	Fontenot et al. (1971a)
	5.49	6.76	Kanev et al. (1971)
	6.48	7.30	Bosman (1973)
	9.30	9.37	Denisov et al. (1973)
	12.50	14.50	Cross and Jenny (1976)
	7.18	7.73	Cullison et al. (1976)
	11.30	9.20	Cross et al. (1978)
Mean	10.18	11.58	

quality were noted. The litter used by Batsman had been dehydrated at 800°C.

Denisov et al. (1973) used up to 40 percent litter waste with growing cattle and reported a resultant depression in daily gain from 0.89 to 0.6 kg, and intake of organic matter/kg gain rose from 9.1 to 12.3 kg. In a subsequent trial the cattle were fed diets containing 0, 15, 25, or 35 percent layer waste. Daily gains were 0.91, 1.02, 1.03, and 0.94 kg and intakes of organic matter/kg gain were 9.5, 8.2, 8.2, and 8.9 kg, respectively. Carcass yield was reported to be higher for the groups given 15 or 25 percent waste.

Meregalli et al. (1973) used ensiled litter with corn forage containing 7.0 percent crude protein, dry-matter basis. Daily gain with and without litter in the silage was 1.32 and 1.23 kg, respectively. Other work on the use of silages containing litter was reported by Creger et al. (1973), Cross and Jenny (1976), Cross et al. (1978), McClure et al. (1979), and Fontenot et al. (1971a). The results suggest that higher levels of litter can be utilized in ruminant diets by means of ensiling than by incorporating dehydrated waste into mixed diets. Fontenot et al. (1971a) reported only an 8 percent drop in daily gain with 25 percent litter in a steer diet, yet Szelenyi et al. (1971) reported a 26 percent drop in gain when 25 percent of a feed mixture was replaced with nonensiled poultry litter. Cross and Jenny (1976) reported no depression in the gain of dairy heifers fed high-corn silage diets containing 0, 15, 30, or 45 percent turkey litter silage, and gain was improved significantly from 0.42 to 0.58 kg/day by the inclusion of 15 percent litter silage in the diet. Although gains were not affected by the level of litter silage in the diet, feed/gain ratio increased from 12.5 at the 0 percent level to 16.7 at the 45 percent level. The results of Cross et al. (1978) tend to confirm their earlier findings (Cross and Jenny, 1976). Steers were fed 30 percent concentrate and 70 percent corn silage, or diets with 0, 10, 30, or 50 percent of the corn silage replaced with broiler litter silage. Daily gain was 0.72, 0.90, 0.94, and 0.63 kg, indicating a significant increase due to the inclusion of litter silage at up to 30 percent and a drop in performance only above that level. Feed/gain ratio showed the same trend and was, respectively, 11.3, 9.9, 9.7 and 10.6. At the 30 percent level, it was estimated that feed costs/kg gain were reduced by 23 percent.

Processed poultry litter has been used in dairy diets. Muftic et al. (1968) reported on the use of broiler litter dried at 60° to 70°C and fed to culled dairy cows in a mixture of 79 percent litter, 20 percent corn, and 1 percent minerals. During 4 months the diet was sufficient for maintaining or increasing body weight of the cows and for maintaining a milk yield of 4 to 6 liters/day. In a later study the same workers found that a similar mixture supplied the requirements for maintenance plus 10 to 20 liters of milk (Muftic et al., 1968). A large experiment by the same group (Muftic et al., 1974) involved two groups of 40 Black Pied cows over 2 years. The experimental group was fed a mixture of 77 percent broiler litter, 20 percent corn, and 3 percent vitamins and minerals, given at the rate of 12 kg daily along with 2 kg hay. Average milk yield with the control and waste diets was 10 and 10.4 kg/day, respectively, and butterfat content of the milk was 4.4 and 3.78 percent. Mean birthweight of calves was heavier with the waste diet, 37.8 versus 36.3 kg. It was reported that there was a higher incidence of reproductive disorders, such as retained placenta, related to the waste diet, possibly because the diet had not been adjusted correctly to ensure fertility. Mello et al. (1973) fed Brown Swiss crossed with Zebu cows on diets containing up to 36 percent poultry litter for 77 days, along with corn meal, corncob meal, and soybean silage. Milk yield or composition was not affected by diet. All diets were well accepted by the cows, although there was some weight loss, which was attributed to the poor quality of the soybean silage.

Galmez et al. (1971a,b) investigated the use of poultry litter in the diet of growing and breeding sheep. Lambs weighing about 28 kg initially were fed for 60 days on diets containing up to 68 percent broiler litter. Daily gain was better with the waste than with alfalfa hay alone. Ewes were fed a control diet or a diet with 63 percent broiler litter for 36 days before and 90 days after lambing. Mean birthweights of single lambs were 4.39 and 4.78 kg and their mean daily gain for 90 days was 146 and 171 g, respectively; the difference was not significant.

Perez-Aleman et al. (1971) reported on the use of dried poultry waste (sterilized broiler litter) in growing swine diets. Levels of 0, 9, 17, or 23 percent were used. Daily gains were 576, 549, 426, and 513 g, and feed/gain ratios were 3.48, 3.72, 4.07, and 4.20, respectively. There was no measurable effect on feed intake.

Blair and Herron (1982) investigated the effects of including 10 percent dehydrated broiler litter in broiler diets. Liveweights at 8 weeks for the control and waste diets were 1,966 and 2,100 g, and feed/gain ratios were 2.14 and 2.13, respectively.

Miscellaneous

Flachowsky and Lohnert (1974) fed adult Merino wethers a control diet or a diet containing 44 percent dried rabbit waste. Intake was not affected, but only the protein was well digested (78 percent).

Cereco protein produced by a patented process was fed at a level of 14 percent in a pelleted diet to rainbow trout (Post and Ward, 1975). No significant difference in gain was noted between the test group and a control group over a 20-week period.

Ochoa et al. (1972) reported on the performance of sheep fed diets containing 0, 10, 20, or 40 percent of a waste mixture (equal proportions of poultry and swine waste). Daily gain was 178, 179, 187, 205, and 175 g, respectively.

PROCESSING

Among the procedures that have been used to process animal wastes prior to feeding are ensiling, dehydration, pelleting, preparation for liquid feeding, oxidation-ditch aerobic processing, commercial (patented) systems, and the use of wastes as substrates for single-cell protein production.

Ensiling

Ensiling is a controlled anaerobic fermentation process during which carbohydrates in the mixture are converted to lactic and other acids. Once sufficient acids are produced, bacterial action ceases and the ensilage is then stable. Heat is generated during the process, and an internal temperature of at least 25°C is usually achieved. Processing wastes by ensiling is economical; it has the further advantage that the process diminishes the hazards from certain potentially pathogenic organisms and renders the waste mixture more palatable.

Feasibility of mixing cattle waste with grass hay and ensiling the mixture was explored by Anthony (1971). The mixture consisted of 57 parts waste and 43 parts grass hay, and the ensiled mixture was termed *wastelage*.

It appears that the addition of animal waste to corn forage before ensiling leads to an improvement in nutritive value. Harmon et al. (1975a) found that the addition of broiler litter to corn forage at 15, 30, or 45 percent of dry matter increased the crude protein content of the silage up to about 18 percent, dry-matter basis. Addition of waste increased final pH values and concentrations of lactic and acetic acids. Dry-matter content of silage was also increased when waste was included at the higher levels.

Residues from medicinal drugs and mineral supplements are probably affected differently by ensiling. Caswell et al. (1978) found that amprolium was present at 10 mg/kg in broiler litter and was unaffected by fermentation. Zinc bacitracin was present at 0.78 units/g in broiler litter and was largely removed by fermentation. Although the broilers had been fed both ethopabate and monensin sodium, no detectable levels were found in the litter either before or after ensiling. Arsenic was present at 6.2 mg/kg in the litter and was not affected by ensiling.

Dehydration

Until the recent escalation in energy costs, dehydration was an attractive processing system. The high temperatures in dehydrators (370° to 700°C) reduced pathogens to low levels and frequently eliminated them entirely. In addition the odor was much reduced or removed. A typical waste dehydrator has been described by Surbrook et al. (1971). Because pathogens are killed in commercial dehydrators, dried poultry waste was the first animal waste product to be accepted by the American Association of Feed Control Officials. Air- or sun-dried waste has been investigated by some workers. This product is more likely to harbor viable organisms because of the relatively low temperatures employed in drying.

One disadvantage of dehydration is that a considerable loss of nitrogen and of other nutrients can occur with dry heat. For instance, Shannon and Brown (1969) reported a 10.6 percent loss in nitrogen at 120°C. Caldertone and Wilson (1976) reported a 20.8 percent loss of nitrogen and a 5.9 percent loss of phosphorus on heating to dryness at 45°C. Silva et al. (1976) reported a 50 percent loss of organic components during drying. Smith and Calvert (1976) reported a 20 percent loss of nitrogen on drying broiler waste. Harmon et al. (1974) have shown that the loss can be reduced by acidification prior to heating. The dehydrated material does not appear to undergo further change on storage for lengthy periods (Chang et al., 1974). This is in contrast to the fresh material as voided: For instance, Feldhofer et al. (1975) reported that crude protein dropped from 34 to 20 percent of dry matter on storage for 21 days. This work indicates that waste as voided (75 percent moisture content in the case of caged layers) should be dehydrated quickly after collection.

At present the main disadvantage of dehydration is the high energy cost. Reduction of moisture content from 75 to 15 percent requires removal of 2,825 kg water/ton dry solids. Additional energy is required to operate the afterburner, found necessary to eliminate odors from flue gases. Esmay et al. (1975) reported that an oil-fired dehydrator and afterburner required 2,220 to 2,770 kcal (9.3 to 11.6 million J) of energy to remove 1 kg water. This indicates a thermal efficiency of 20 to 25 percent. Muller (1976) estimated the cost of dehydrating at $25–$50 per ton of dehydrated material.

Shannon et al. (1973) conducted a bacteriological survey on eight samples of commercial dehydrated poultry waste used for animal feeding. The samples were not sterile and the organisms found were *anthracoid* bacilli, *paracolon* bacilli, *Staphylococci*, and *E. coli.*

When voided, layer waste contains about 75 percent water (and generally more if excessive water spillage occurs). An obvious solution to the reduction in dehydration costs is to predry the material before dehydration. Bressler and Bergman (1971) developed a two-stage drying system using mechanical stirring of the waste in the storage pits and high-velocity fans to remove up to 83 percent of the moisture before the waste was put into the dryer. Ostrander (1975) described a high-rise layer battery design in which waste is allowed to build up in the form of cones on slats under flat-deck cages. A design involving a drying tunnel in the barn was described by Sheppard et al. (1975). Waste predryed in this way may contain 14 to 40 percent moisture (Blair and Herron, 1982; Ostrander, 1975) and would be much less costly to dehydrate.

It can be concluded that dehydration results in a product that can be used successfully in livestock feeding, but the process may not be economically feasible due to high energy costs.

Other Processes

Pelleting

Pelleting animal wastes prior to feeding was investigated by Hull and Dobie (1973), Smith and Lindahl (1977), and Smith et al. (1976). This system has the advantage that ingredient-sorting by animals is prevented; in addition the heat of pelleting is probably beneficial. However, a disadvantage is that the waste has to be dried or dehydrated before it will pellet successfully.

Liquid Feeding

A liquid feeding system for layer waste was investigated for ruminants by Evans et al.(1978a,b) and Smith et al. (1978). The addition of 2 percent molasses or 1 percent propionate was found to increase feed intake when the waste was fed along with corn silage. Intakes of nitrogen and minerals were sufficiently high with this system, suggesting that it merits further attention. One probable disadvantage of this type of system is its potential for transmitting disease. Smith et al. (1978) reported that 38 percent of the samples of waste were positive for salmonellae.

Oxidation Ditch

Aerobic digestion of swine waste using an oxidation ditch has been described by Day and Harmon (1974) to produce a nutrient-rich drinking water. As a result of single-cell protein production in the waste liquor during the digestion process, the protein content of the diet fed to swine could be reduced by 15 percent. This system should be directed toward nonruminants because they cannot utilize the nonprotein nitrogen present in the waste unless converted into protein through the mediation of microorganisms. Several potential health problems exist with this system. Buildup of intestinal worm eggs has been noted, and there can be an increase in nitrate concentration in the ditch liquor. The survival time of *Salmonella typhimurium* in a model oxidation ditch was 17 days at summer temperatures and 47 days at winter temperatures (Will et al., 1973).

Patented Systems

Several commercial (patented) systems have been developed for processing animal wastes for feeding. Techniques involved include ensiling and fractionation (Cereco process) and chemical treatment (Grazon and Corral systems).

Substrates for Protein Production

The use of wastes as substrates for protein production has been described by Calvert (1976). Among the organisms grown on these substrates were algae, yeasts, fungi, bacteria, house-fly larvae, and earthworms. According to Calvert (1979), none of the systems involving algae, yeasts, bacteria, fungi, insects, or earthworms contribute greatly to the supply of protein supplements for livestock feeds. Algal systems have been tested more adequately and appear to yield the most promising results. In the algal systems the amount of nitrogen converted to protein is as great or greater than that of any of the other systems, and the protein quality is high. Systems using yeasts, bacteria, and fungi all appear to show promise, but little published data critically evaluating the systems are available. Yields of protein were low when insects and earthworms were grown on waste as a substrate. These systems probably require a greater degree of technology than other systems, which may preclude their adoption as on-farm systems.

UTILIZATION SYSTEMS

Experimental

A number of systems involving algae, yeasts, bacteria, fungi, insects, and earthworms for the conversion of animal wastes were reviewed recently by Calvert (1979). It was concluded that the algal systems had produced the most promising results. The algal system developed by Dugan et al. (1969, 1971) has the advantage that methane is produced during the process. In addition, it is claimed that after algal separation the water can be recycled. Potential algal yield in this system was estimated at 11 to 15 tons dry matter/ha/year. The system developed by Miner et al. (1975), which uses swine waste, was projected to yield 121.5 tons dry matter/ha/year. Work is required to determine adequately the nutritional value of algae (*Chlorella*), but results to date suggest that algal protein may be comparable with protein from some conventional sources.

The commercial feasibility of growing feed yeasts on hydrocarbon substrates suggests that work needs to be carried out on the use of animal wastes as substrates for yeast growth. Little work appears to have been done in this area (Calvert, 1979).

In some systems bacterial cultures are used to digest animal wastes, and the digested material is of potential value as feed. The processed solid material has been fed successfully to cattle (Vetter, 1972) and to swine (Harmon et al., 1972) but problems appear to exist with the liquid (oxidation-ditch mixed liquor). Johnson et al. (1977) supplied this liquor as drinking water to laying hens and reported a rapid drop in egg production from 64.7 to 1 percent. The loss was attributed to increases in the levels of dissolved oxygen and nitrate from 4.7 to 6 and 210 to 1,300 mg/kg, respectively, suggesting that monitoring of these and perhaps other constituents would be necessary.

Umstadter (1980) described a unique system for the treatment and utilization of cattle waste. Lagoons were used to purify the liquid waste and promote algal growth. Subsequently, *Tilapia* fish were introduced to utilize the algae (the fish could then be harvested). The sludge remaining after methane production was stated to be acceptable by cattle as feed, which could possibly replace 50 percent of the protein supplement in the diet.

Experimental systems involving the treatment of animal wastes for liquid feeding have been investigated by Evans et al. (1978b) and Smith et al. (1978).

Other systems have involved ensiling alone or with other feeds. For instance, Harmon et al. (1975b) found that the nutritive value of corn silage was improved by the addition of broiler litter. Acceptability of waste was also improved by ensiling. Anthony (1969) reported satisfactory results from feeding ensiled cattle waste and grass hay. Saylor and Long (1974) showed that a satisfactory silage could be made from 60 percent poultry waste and 40 percent grass hay.

Industrial

Successful industrial systems involving dehydration, ensiling, liquid-solid separation, and chemical treatment have been developed.

Dehydration involves passing the raw waste through a heat chamber, which removes volatile material and exhausts it with the flue gases. As a result, an afterburner is now generally added to most dehydrators. A wide variety of dehydrators are in use commercially, and considerable use, especially, is made of dehydrated poultry waste in animal feeding.

Ensiling is used extensively for feeding wastes to livestock and has the advantage that it is an on-farm system. A further advantage is that it is inexpensive. One disadvantage is that it may be seasonal in use.

Various liquid-solid separation systems are being used to process wastes (Ward et al., 1975). These systems have the advantage that they are mechanized and that the processed fractions have good animal acceptability. However, they are presently justified only in large feeding operations.

Chemical treatment of poultry wastes for incorporation into cattle diets is being practiced in the southern United States (Masters, 1977). Advantages of this system are that no storage is required and that the treated waste can be used directly after treatment.

Potential Utilization

Animal wastes represent a very large but almost untapped resource for use as animal feed. Methods of utilizing this resource will vary, and they depend on a variety of factors. For example, minimum processing will be needed for poultry litter or cattle waste from the dry areas of the country, but considerable processing will be required for materials such as cattle waste from the humid east, swine waste, and caged layer waste. Because of their relatively low energy and their high level of nonprotein nitrogen, some of these wastes are best suited for use as feed for ruminants. Such high protein wastes as those from poultry production would be best suited for use in limited quantities as protein and mineral supplements. None of the wastes is high in available energy. The low-protein wastes will likely be used in substantial quantities in low-producing animals or in limited amounts in high producers.

A variety of technically feasible processing methods is available. Unfortunately, not all of these are economically feasible. Some of the more important processing methods are dehydration, ensiling, solid-liquid separation, chemical treatment, and autoclave. It is likely that the methods that have a low requirement for fossil fuel will be the methods of choice. The most feasible processing method appears to be ensiling, alone or in combination with other ingredients. Combining high-moisture waste with low-moisture ingredients and low-moisture waste with high-moisture ingredients offers the most potential for high-quality silage. For example, caged waste could be ensiled with crop residues, and broiler litter with wastes from fruit or vegetable processing.

The value of animal wastes will depend on their nutritional value and the price of other feeds. The wastes have been shown to be quite valuable when compared with other feedstuffs (Smith and Wheeler, 1979). It is also likely that the animal wastes will be used for production of fossil fuels, such as methane, ethanol, and fatty acids, and the residue used as feedstuffs. In fact, these energy-generating processes probably will not be economically feasible unless considerable value is recovered in the residue as a feedstuff.

ANIMAL AND HUMAN HEALTH

Human and animal health aspects of feeding livestock wastes have been reviewed by Fontenot and Webb (1975) and McCaskey and Anthony (1979).

Pathogenic Organisms

There is no doubt that raw animal waste may contain pathogenic organisms. However, adequate processing renders the waste free of pathogens or with a much reduced profile of organisms capable of causing disease. One documented disease outbreak has apparently been linked to the feeding of animal waste, and it was attributed to faulty processing. Egyed et al. (1978a,b) described a disease outbreak in over 1,000 cattle and sheep in Israel, involving 25 farms in which feed containing 10 percent poultry litter was used. A diagnosis of botulism was suggested (Cohen and Tamarin, 1978), although the toxin and organism could only be isolated from one animal. The botulism organism (Type D) appears to be endemic to Israel, since outbreaks have occurred in animals fed other feeds (Fontenot and Jurubescu, 1980). Total bacterial colonies were significantly increased in the silages containing 45 percent waste, but coliforms were not significantly different nor significantly lower in litter silage than in unsupplemented forage silage.

No disease problems have been reported with poultry wastes in practical diets for beef cattle, dairy cattle, or sheep (Bucholtz et al., 1971; Bull and Reid, 1971; Drake et al., 1965; El-Sabban et al., 1970; Fontenot et al., 1966; Johnson et al., 1975; Liebholz, 1969; Noland et al., 1955; Southwell et al., 1958). Even calves, which are well known to be susceptible to digestive upsets, remained healthy when fed diets containing wet cage layer waste (Smith et al., 1978). Feedlot cattle also remained healthy when fed diets containing cage layer waste treated with organic acids (O. B. Smith et al., 1979), and the researchers concluded that potential health problems were no more serious than with conventional feeds. Similarly, no disease problems were encountered when cattle waste was fed to cattle and poultry (Anthony, 1966, 1971; Durham et al., 1966), cage layer waste to layers (Flegal and Zindel, 1971), or swine waste to swine (Harmon, 1974).

Nevertheless, animal wastes commonly contain pathogenic organisms (Alexander et al., 1968; Caldertone and Wilson, 1976; Carriere et al., 1968; Caswell et al., 1978; Knight et al., 1977; Kraft et al., 1969; O. B. Smith et al., 1978, 1979) and should be processed before being fed.

Alexander et al. (1968) examined 44 field samples of poultry litter and found that 13 were negative for pathogenic bacteria. The other samples tested positive for 10 different species of *Clostridium*; 2 of *Cornebacterium*; 3 types of *Salmonella*; and various actinobacilli, *Mycobacteria, Enterobacteriaceae, Bacilli, Staphylococci, Streptococci*, and yeasts (Table 42). Three *Clostridia* were regarded as pathogenic types (*C. chauvoei, novyi*, and *perfringens*) but were not harmful when injected intramuscularly into guinea pigs. Many of the organisms found were regarded as

TABLE 42 Results of Bacteriological Analysis of 44 Samples of Poultry Litter

Types Isolated	Number Isolated
Clostridium perfringens	8
Clostridium chauvoei	1
Clostridium novyi	8
Clostridium sordellii	1
Clostridium butyricum	2
Clostridium cochlearium	1
Clostridium multifermentans	1
Clostridium carnis	1
Clostridium tetanomorphum	1
Clostridium histolyticum	1
Corynebacterium pyogenes	1
Corynebacterium equi	2
Salmonella blockley	1
Salmonella saint-paul	1
Salmonella typhimurium var. Copenhagen	1
Actinobacillus spp.	1
Yeast	1
Mycobacterium spp.	2
Enterobacteriaceae (other than *Salmonellae*)	All samples
Bacillus spp.	All samples
Staphylococcus spp.	All samples
Streptococcus spp.	All samples

SOURCE: Alexander et al. (1968).

normal inhabitants of the intestinal tract of animals. Litter that was negative for *Salmonella choleraesuis* was inoculated with this serotype and remained contaminated for 29 days. Carriere et al. (1968) reported finding mycobacteria in 8 out of 29 samples of poultry litter. They concluded that cattle fed poultry litter might be infected or sensitized to *Mycobacterium avium* and/or atypical types of mycobacteria, leading to false positive reactions and incorrect diagnosis of tuberculosis. Kraft et al. (1969) tested composite samples of freshly voided excreta from 91 poultry houses and found that 29 percent were positive for salmonellae. The houses were located on 36 farms, 18 of which yielded one or more positive samples. Smith et al. (1978) found that the rate of *Salmonellae* contamination in fresh layer waste was 38 percent. Diets containing waste were also found to be contaminated.

Fungi also may be present in animal and poultry wastes (Lovett, 1972; Lovett et al., 1971; Singh, 1974). However, they have been isolated from poultry feeds also (Lovett, 1972; Singh, 1974). The types found in litter by Lovett (1972) and Lovett et al. (1971) were *Penicillium, Scopulariopsis,* and *Candida,* mainly, and in feeds were mainly *Penicillium, Aspergillus, Fusarium,* and *Mucor.*

The data indicate that processing has a marked pasteurizing effect. Harmon et al. (1975a) harvested corn forage at 30 or 40 percent dry matter and ensiled it with litter from broilers kept on wood shavings to supply 15, 30, or 45 percent of the dry matter. Total bacterial counts/g were significantly increased in the silages containing 45 percent waste, but coliforms were not significantly increased or were significantly decreased in the litter silages, compared with the control silage. McCaskey and Anthony (1975) investigated the survival of organisms in a mixture of 45 parts ground shelled corn, 15 parts corn silage, and 40 parts cattle waste, which was blended and then ensiled. A total of 27 salmonellae types was added to the mixture prior to ensiling to obtain recovery data. Ninety-two percent were recovered in the fresh waste when tests were carried out prior to ensiling. None were recovered after ensiling three days at 25°C (see Table 43). After ensiling for 4 days, 78 percent were recovered when the temperature was 5°C and 93 percent were recovered when the temperature was 15°C. When the ensiling temperature was 25°C, 4 percent were recovered; none was recovered when the temperature was 35°C. The minimum acid level for the growth of salmonellae was found to be pH 4.6 to 5.0. At pH 4, growth was completely inhibited. These workers also investigated the effect of a 4-day ensiling period on the growth of a variety of organisms at increasing temperatures. Yeasts and molds were greatly reduced, coliforms eliminated, and acid-producing bacteria increased by low temperatures then decreased greatly at the higher temperatures (see Table 44). Spore-forming bacteria were fairly stable in number regardless of temperature, although anaerobic spore-formers were reduced

TABLE 43 Survival of *Salmonellae* in Cattle Waste and in an Ensiled Waste-Feed Mixture[a]

	Waste	Mixture
Percentage recovery	92	0
pH, initial	6.8	6.5
pH, 3 days	6.0	4.5

[a]Recovery after 3 days incubation at 25°C.

SOURCE: McCaskey and Anthony (1975).

TABLE 44 Effect of Temperature on Survival of *Salmonellae* in an Ensiled Waste-Feed Mixture

	Temperature (°C)			
Survival After 4 Days Ensiling	5	15	25	35
Number of cultures[a]	21	25	1	0
Percent of cultures	78	93	4	0
pH, initial	4.8	4.8	4.8	4.8
pH, 4 days	4.6	4.4	4.0	4.1

[a] 27 salmonella cultures were used.

SOURCE: McCaskey and Anthony (1975).

at higher temperatures (15°C or above). At 5°C the numbers of aerobic spore-formers increased.

Knight et al. (1977) investigated microbial population changes and fermentation characteristics of ensiled, bovine-waste blended diets. The mixtures contained up to 60 percent waste and were ensiled at 25°C for 10 days. Coliforms were eliminated in mixtures with 40 and 60 percent waste after 5 days of ensiling and after 10 days in the 20 percent mixture. *Salmonellae* were isolated twice prior to ensiling but were not isolated after 3 days of ensiling. Spore-forming bacteria survived but did not proliferate in the ensiled mixtures. Reduction in bacterial contamination was related to a reduction in pH to around 4.5. Yeasts and molds were reduced from around 10^6/g to around 10^4/g after ensiling for 10 days. Caswell et al. (1978) found that coliform bacteria in broiler litter were eliminated by ensiling at 25 to 50 percent moisture. *Proteus* organisms were destroyed by ensiling at all moisture levels tested. These studies indicate that the beneficial effects of ensiling can be attributed to two causes, heating and the development of acid. Ensiling also has a beneficial effect on contamination with parasitic nematode larvae. Ciordia and Anthony (1969) found that feedlot waste contained viable larvae, but none was present in the silage made from waste and dry hay. Farquar et al. (1979) reported that sporulation of bovine coccidia was prevented by ensiling a waste-blended diet.

A number of reports indicate the beneficial effects of heat processing. Fontenot et al. (1970) found no adverse effects in sheep fed for 80 days on diets containing up to 75 percent poultry litter that had been sterilized by dry heat at 150°C for 4 hours. Parameters included physiological observations during the growth period, detailed necropsies, and histological investigations. Johnson et al. (1975) reported a 91-day test involving

24 yearling beef calves fed diets containing 0, 10, or 15 percent dehydrated feedlot waste. All animals were necropsied. Gross and histological examinations of rumen, abomasum, small intestine, liver, lung, kidney, trachea, heart, and gall bladder did not indicate any causal relationship between incidence of pathological lesions and waste feeding. Shannon et al. (1973) reported on the bacteriological status of 8 samples of dehydrated poultry waste. None was sterile, but the numbers of organisms were low. Found were *Anthracoid* and *Paracolon* bacilli, *Staphylococci,* and *Escherischia coli. Salmonellae* were absent. Caswell et al. (1975) tested four methods of processing broiler litter that was heavily contaminated with bacteria:

1. Dry heating litter 0.6 cm deep at 150°C for 10 to 20 minutes
2. Autoclaving at 5.0 cm deep at 121°C and 105 kg/cm^2 pressure for 5 to 30 minutes
3. Dry heating at 150°C for 15 minutes at depths of 0.6 or 2.5 cm after addition of paraformaldehyde at 0 to 5 g/100 g litter
4. Fumigation with ethylene oxide at 22°C at 1 atm for 30 to 120 minutes

They found that coliforms were eliminated by all treatments except 30 minutes fumigation. Total bacterial counts were acceptable (20,000/g) after treatment 1 for 20 minutes, treatment 2 for 10 minutes, treatment 3 with 1 to 4 g paraformaldehyde/100 g litter, and treatment 4 for over 30 minutes.

Chang et al. (1975) reported the results of a microbiological survey (see Table 45) on dehydrated poultry waste. They found an inverse relationship between the temperature of the dehydrator and the number of organisms and between the moisture content of the dehydrated waste and the number of organisms (see Table 46). Below 11 percent moisture the number of organisms was reduced drastically. Only four groups of bacteria were recovered when the dehydration temperature was set at 260°C or above. Blair and Herron (1982) reported that dehydrated broiler litter and layer wastes showed low to scant contamination with *E. coli,* and no salmonellae.

Under the California quality control standards for processed waste products, the criteria for effectiveness of pasteurization require not more than 20,000 bacteria and 10 coliform organisms/g dried product, and freedom from salmonellae (Helmer, 1980).

In assessing the risk of pathogenic organisms associated with processed wastes, it is fair to point out that regular feed ingredients are commonly contaminated. Singh (1974) found that all ingredients tested contained a minimum total microbial count of 12×10^3/g and a mold count of at least 3×10^3/g. An average incidence of 4 to 5 percent salmonella contamination of various animal feeds was reported by Allred et al. (1967) and Hauge and Bovre (1958), and outbreaks of salmonellosis in farm stock have been traced to feed and feed ingredients (Boyer et al., 1958, 1962; McClarin et al., 1959; Pomeroy and Grady, 1960). Other possible health hazards associated with feed ingredients include molds and mold toxins (McCaskey and Anthony, 1979).

It may be concluded that the health risks from pathogenic organisms associated with the feeding of adequately processed animal wastes are probably no greater than those associated with the feeding of meat meal tankage, blood meal, poultry by-products meal, hydrolyzed poultry feathers, offal meal, or processed paunch product, all of which are approved for use in feed.

Harmful Substances

Minerals

The mineral content of animal wastes could lead to at least two potential problems, toxicity and accumulation in tissues or in the environment. However, the data suggest that only the copper content of wastes fed to sheep is of real concern (Fontenot et al., 1971b). This could be anticipated since it is well known that sheep are very sensitive to the copper level of

TABLE 45 Microorganisms Recovered From Samples of Poultry Waste

Microorganisms	Note
Aerobacter aerogenes	
Alkaligenes faecales	
Bacillus spp.	a
Clostridium spp.	a
Corynebacterium spp.	
Enterobacter spp.	
Escherichia coli	b
Lactobacillus spp.	a
Proteus spp.	
Streptococcus spp., fecal	a

[a]Bacteria recovered only when dehydration temperature was 260°C or higher.
[b]Recovered only in high moisture samples.

SOURCE: Chang et al. (1975).

TABLE 46 Effect of Temperature and Moisture on Microbial Counts of Dehydrated Poultry Waste

Microbes	Dehydration Temperature	Sample Moisture (%)	Average Microbial Count/g
Aerobic	260°C	>10	20,281,666
		<10	710,000
	Over 260°C	>10	6,719,520
		<10	183,396
Anaerobic	260°C	>10	6,958,333
		<10	730,000
	Over 260°C	>10	1,360,530
		<10	46,241

SOURCE: Chang et al. (1975).

the diet. Other elements of concern are mercury, cadmium, lead, arsenic, and selenium.

The copper content of animal wastes will vary with the amount added to the diet of the host animal. For instance, copper may be added in the form of copper sulfate as a mold-control agent, in addition to its requirements as an essential nutrient. The analytical data presented in Appendix Table 4 suggest that the mean content of copper in cattle, swine, and poultry wastes (dry-matter basis) is, respectively, 31.0, 114.2, and 70.3 mg/kg. The range is probably considerable, being 62.8 to 249 mg/kg for swine waste and 31.4 to 300 mg/kg for caged layer waste, according to the values used to derive the means shown in Table 4. Blair (1974) quoted a range of 28 to 109 mg/kg dry matter in dehydrated layer waste. Berryman (1970) reported 675 mg copper/kg dry matter in slurry from swine units in the United Kingdom, where copper was used as a growth stimulant.

The dietary level of copper that can be expected to cause problems with sheep is around 15 mg/kg dry matter (Suttle and Price, 1976). Most of the copper in dehydrated waste has been shown to be chemically available; therefore the recycling of animal wastes through sheep constitutes a real hazard unless steps are taken to keep the total dietary copper content at a safe level for this species. Increasing the sulfur and/or molybdenum contents of the diet could be worthwhile to reduce the uptake of copper (Suttle and Price, 1976). Thomas et al. (1972) found that the inclusion of 0, 25, or 50 percent dehydrated poultry waste in diets for growing sheep resulted in normal copper values in the liver and kidney.

Other species of farm livestock appear to be much less sensitive to dietary copper level. Webb et al. (1979) wintered cows over 4 years on a diet of 80 percent broiler litter containing up to 160 mg/kg added copper and found that performance was unaffected although liver copper value increased from 58.8 mg/kg with a control diet to 561.3 mg/kg with the test diets. They also found that liver copper decreased during the summer when the cows were not receiving litter.

Elements other than copper are probably of much less significance in terms of likely hazards in any recycling system. Westing and Brandenberg (1974) fed steers for 184 days on a control diet containing 2.6 mg/kg lead or on a diet containing 14 percent composted feedlot waste that contained 3.6 mg/kg lead. Concentrations of cadmium and lead in the liver of control and test animals were, respectively, 0.062 and 0.567 mg/kg, and 0.041 and 0.460 mg/kg.

Thomas et al. (1972) reported that calcium, phosphorus, sodium, potassium, magnesium, zinc, iron, copper, and manganese levels in liver and kidney of sheep fed diets containing 0, 25, or 50 percent dehydrated poultry waste were within a normal range.

Webb and Fontenot (1975) reported that the inclusion of 25 or 50 percent broiler litter containing copper at a concentration of 230 mg/kg tended to increase liver copper in finishing steers after a 5-day withdrawal period. In a subsequent trial, the feeding of diets with litter containing copper at a concentration of 289 mg/kg increased the level of copper in the liver. Muscle copper tended to be higher in the cattle fed broiler litter in both trials. Westing et al. (1977) fed diets based on 70 percent ensiled corn and 30 percent broiler litter (dry-matter basis) to fattening heifers. Concentrations of bromine, arsenic, cadmium, copper, mercury, molybdenum, vanadium, and zinc were higher in the litter than in the other feedstuffs. Only liver copper was increased in the cattle fed the corn-litter silage. None of the mineral concentrations were elevated in muscle. Calvert and Smith (1976) conducted a trial in which steers were fed a control diet or a diet containing 12.1 percent dehydrated poultry waste for 400 days. The concentrations of various mineral elements in the test and control diets, respectively, were copper, 17.06 and 4.90 mg/kg; zinc, 62.31 and 30.11 mg/kg; iron, 458.4 and 127.6 mg/kg; cadmium, 0.082 and 0.092 mg/kg; and lead, 0.18 and 0.34 mg/kg. Mineral concentrations in the tissues of animals fed the test and control diets were, respectively, liver copper, 332.5 and 157.7 mg/kg; kidney iron, 261.4 and 243.4 mg/kg; and kidney cadmium, 1.7 and 5.0 mg/kg (dry-matter basis).

Several organic arsenicals have been approved for use in poultry and swine diets; therefore attention has to be given to possible arsenical residues. The arsenic content of wastes may range from 0.43 to 44 mg/kg (see Appendix Table 5).

The current food standards allow up to 1 or 2 mg/kg arsenic in meats, which is much higher than any level found in tissue following the inclusion of waste in the diet. El-Sabban et al. (1970) found that the inclusion of poultry litter containing 17 mg/kg arsenic in the diet of sheep resulted in a significant increase in the arsenic content of liver, but not above 0.015 mg/kg. The presence of arsenical residues has been reported in the milk and blood of cows (Calvert and Smith, 1972) and in the tissues of cattle (Webb and Fontenot, 1975) and sheep (Calvert, 1973, 1975) following the feeding of poultry litter. The levels reported by Webb and Fontenot (1975) after a 5-day withdrawal were 0.2 mg/kg (dry-matter basis) at the highest.

Polidori et al. (1972) reported on the use of dehydrated layer waste containing 0.98 mg/kg arsenic (dry-matter basis). When included in the diet of laying hens at a level of 10 percent, the resultant eggs had 0.32 and 0.37 mg arsenic/kg and control eggs had 0.28 mg/kg, but the difference was not significant.

Calvert (1973) studied the retention of arsenic in sheep fed diets containing 0, 7, or 14 percent dried broiler waste. The broiler diet contained 3-nitro-4-hydroxyphenyl arsonic acid at 50 mg/kg and resulted in an arsenic concentration of 42 mg/kg in the waste. About 87 percent of the arsenic was excreted, mainly (76 percent) in the feces. About 2.4 mg arsenic was retained by the sheep, resulting in a tissue concentration of about 0.08 mg/kg. When sheep were given diets containing up to 300 mg/kg arsenic they accumulated amounts proportional to the amounts ingested, mainly in the liver and kidney, though no toxicity symptoms were noted. Concentrations in blood, liver, kidney, urine, and feces fell rapidly after withdrawal of the mineral.

Blair and Herron (1982) included poultry waste containing 4 to 31 mg/kg arsenic at a 10 percent level in broiler diets and recorded the arsenic in liver, leg muscle, and breast muscle. The highest level recorded was 0.25 mg/kg (fresh-weight basis), and there was no correlation between tissue level and treatment. Withdrawal had no obvious effect on arsenic concentration in tissue, possibly because of the low concentrations found. Calvert (1973), however, reported that a withdrawal period of 5 days was efficacious after animals had been fed a high level of arsenic in the diet (273.3 mg/kg as arsanilic acid for 28 days). The concentration of arsenic with no withdrawal was 29.2 mg/kg (dry-weight basis) and 5.0 mg/kg following withdrawal. A withdrawal period of 7 days was required with dairy cows.

Vijchulata et al. (1980) reported that tissue mineral levels were altered somewhat in steers fed diets containing up to 25 percent cage layer waste. Copper level in liver increased from 155 to 490 mg/kg and magnesium and phosphorus in kidney increased from 794 to 827 and 8,615 to 9,519 mg/kg (dry-matter basis), respectively. Arsenic level in kidney and muscle increased from 0.11 to 0.33 and 0.1 to 0.25 mg/kg (dry-matter basis), respectively, but the levels remained below the safety level of 2 mg/kg in liver and 0.5 mg/kg in edible meat established by the Food and Drug Administration.

Medicinal Drug Residues and Metabolites

Various drugs are used in animal production for medicinal purposes and to improve growth and feed efficiency. Many require a withdrawal period before slaughter to avoid harmful residues in the carcass. It is reasonable to expect that during the period of feeding these drugs will be excreted in the feces and/or urine. Elmund et al. (1971) reported that up to 75 percent of the chlortetracycline in the diet of beef animals was excreted. Bacitracin is not absorbed, and therefore none is expected to appear in edible tissues.

According to Donoho (1975), 75 percent of the monensin fed to steers is excreted in the feces as the parent compound. Dehydrated waste from poultry fed the compound contained monensin at a concentration of 10 to 15 mg/kg.

Bevill et al. (1978) found that feces and urine from swine fed a diet containing sulfamethazine at a concentration of 100 mg/kg served as a source of the drug for other animals having access to the excreta. Concentrations of sulfamethazine in the blood of swine having access to the excreta were of sufficient magnitude to result in tissue concentrations exceeding 0.1 mg/kg, the maximum amount allowed in pork according to the Feed and Drug regulations. Zero tolerance of sulfamethazine in pork is allowed under the Canadian regulations.

Only limited research has been conducted on medicinal drug residues in animal waste and in the edible products of animals fed the waste. El-Sabban et al. (1970) reported that in steers fed processed poultry waste, chlorinated hydrocarbon compounds in back-fat and arsenic in liver were found in amounts of less than 1 mg/kg. Cregar et al. (1973) reported findings with heifers fed a silage based on broiler wood shavings litter. The birds had been fed diets containing amprolium and ethopabate as coccidiostats, and zinc bacitracin and 3-nitro-4-hydroxyphenyl arsonic acid as growth promoters. Zinc bacitracin at 1.53 and arsenic at 68.52 mg/kg dry matter were detected in the silage, but the other drugs were not detected. No residue of any drug was found in muscle, liver, or fat.

Furazolidone levels of 10.2 to 25.1 mg/kg and nitrofurazone levels from 4.5 to 26.7 mg/kg were reported in samples of poultry litter by Messer et al. (1971). Levels of drugs in broiler litter from Virginia are shown in Table 47 (Webb and Fontenot, 1975).

No residues of amprolium or arsenic were detected in the heart, spleen, 12th rib (edible tissue), kidney, kidney fat, liver, or brain of lambs fed poultry litter containing amprolium and 3-nitro-phenyl arsonic acid with

TABLE 47 Drug Residues in Broiler Litter

Drug	Concentration[a] Average	Range	Number of Samples
Oxytetracycline (mg/kg)	10.9	5.5 – 29.1	12
Chlortetracycline (mg/kg)[b]	12.5	0.8 – 26.3	26
Chlortetracycline (mg/kg)[c]	0.75	0.1 – 2.8	19
Penicillin (units/g)	12.5	0 – 25.0	2
Neomycin (mg/kg)	0	0	12
Zinc bacitracin (units/g)[d]	7.2	0.8 – 36.0	6
Zinc bacitracin (units/g)[e]	12.3	0.16– 36.0	5
Amprolium (mg/kg)	27.3	0 – 77.0	29
Nicarbazin (mg/kg)	81.2	35.1 –152.1	25

[a] Dry-matter basis.
[b] Chlortetracycline used continuously in broiler diets.
[c] Chlortetracycline used intermittently in broiler diets.
[d] Zinc bacitracin used in broiler diets.
[e] Zinc bacitracin not used in broiler diets.

SOURCE: Webb and Fontenot (1975).

and without additional drugs (Brugman et al., 1967). Webb and Fontenot (1975) investigated tissue levels of nicarbazin, amprolium, and chlortetracycline in finishing cattle fed diets with 0, 25, and 50 percent broiler litter after a 5-day withdrawal. Chlortetracycline was detected at an average concentration of 41 mg/g in kidney fat from two steers fed 50 percent litter and at a concentration of 34 mg/g in kidney fat from one steer fed 25 percent litter. The chlortetracycline content of the litter was 12.5 mg/kg dry matter. No chlortetracycline was detected in kidney fat from four other steers fed 50 percent litter and from five other steers fed 25 percent litter. No residues of nicarbazin or amprolium were found in any of the tissues of 24 cattle fed litter containing amprolium at a concentration of 42.3 or 51.3 mg/kg.

Helmer (1980) reported that monitoring of processed animal wastes in California has suggested that drug residues have not yet posed a problem.

These data suggest that drug residues in tissues of animals fed wastes are very low, provided a withdrawal period is allowed before slaughter.

Other

Mycotoxins are metabolites of fungi and are produced by a variety of species. Many have been found in animal feeds due to the presence of fungi. They are now known to be of importance since they can cause problems in poultry and livestock. Some are also known to be carcinogenic.

Little research has been done on the occurrence or formation of mycotoxins in animal wastes, although formation of aflatoxins under laboratory conditions was demonstrated by Hendrickson and Grant (1971). Lovett (1972) suggested that poultry litter may be no more of a problem than feed. Blair and Herron (1982) tested dehydrated broiler litter, in-house dried layer waste, and dehydrated layer waste for mycotoxins and were unable to detect aflatoxin, ochratoxin, or zearalenone.

Hesseltine (1976) advocated the prevention of mycotoxin formation in foods and feeds rather than detoxification once the toxins had been formed. It would seem sensible to apply the same principle to animal wastes. Since the fungi do not grow on dry substrates, the wastes should be dehydrated rapidly after collection. Another reason for advocating rapid dehydration is that *Aspergillus flavus* does not produce aflatoxins for 48 hours after spore germination under the most favorable conditions.

Residues from chlorinated hydrocarbon pesticide residues and industrial contaminants do not appear to be a problem with the feeding of dehydrated poultry waste according to Smith et al. (1976). Chickens were given feed containing 20 mg/kg PCBs, which was 100 times the U.S. Food and Drug Administration guideline of 0.2 mg/kg for complete animal feeds. When the dehydrated poultry waste was included in a dairy concentrate at 32 percent, the highest PCB residue found in milk fat was 5 mg/kg, which was only twice the guideline of 2.5 mg/kg. Residues dropped rapidly within the first 10 to 15 days after PCB feeding stopped. The behavior of PCBs in the animal is regarded as being similar to that of chlorinated hydrocarbon pesticides and industrial contaminants.

A high incidence of abortion was reported in cows that were fed low levels of poultry litter in the wintering ration and were subsequently grazed in the summer on a pasture that had been fertilized with litter (Griel et al., 1969). The cause of the problem was not established. The litter contained estrogenic activity from feeding dienestrol acetate to the birds, a practice no longer approved. The authors suggested that a hormone imbalance was involved, but they pointed out that use of diethylstilbestrol in previous work at higher levels than the estrogenic residues in the waste had not caused abortion. Some producers have been feeding poultry litter to their cow herds for more than 10 years with no abortion problem (Council for Agricultural Science and Technology, 1978).

Quality of Products from Animals Fed Waste

A very important question is whether the quality of food products is affected by feeding waste. This has been investigated by a number of workers, some of whom did not use a withdrawal period in their investigations.

No differences in meat quality as judged by chemical analysis or by studies of tenderness, juiciness, or flavor of meat from beef animals fed diets containing up to 50 percent waste were reported by Andersen et al. (1976), Cross et al. (1978), Fontenot et al. (1971a), Kanev et al. (1971), Vijchulata et al. (1980), and Ward et al. (1975). Cregar et al. (1973) reported that a taste panel of 50 people judged steaks from animals fed broiler litter silage at an average intake of 5.5 kg/day to be slightly less acceptable than steaks from control animals on the basis of tenderness, flavor, and juiciness. This might have been related to a lack of finish in the waste-fed animals. Rhodes (1972) concluded that the tenderness or juiciness of meat from beef animals was not significantly affected by waste feeding, although the flavor of beef from animals fed dehydrated layer waste was judged as being significantly poorer than that of animals fed a control diet or a diet containing litter waste. Five instances of off-flavors were noted but they were not related to treatment.

Results with dairy cows fed diets containing up to 36 percent poultry waste suggest no significant effects on milk quality as judged by composition and flavor (Denisov et al., 1975a; Kristensen et al., 1976; Mello et al., 1973; Silva et al., 1976). Kristensen et al. (1976) reported that after 14 days storage, the milk from cows fed waste had a stronger flavor than control milk. Waste feeding did not affect rennet coagulation of milk, the process of acidification, or the quality of cheese made from it.

No deleterious effects as a result of waste feeding have been reported on egg quality (Faruga et al., 1974; Flegal and Zindel, 1971; Kienholz et al., 1975; Vogt, 1973). Biely et al. (1972) reported in one trial that with 20 or 30 percent waste in the diet, eggs had a slightly reduced albumen quality as judged by Haugh unit score. Lee and Bolton (1977) reported no effect on albumen quality or on incidence of shell cracking, but shell weight and shell thickness were poorer with dehydrated layer waste in the diet. Probably this effect could be related to an inadequate level of calcium in the diet.

REGULATORY ASPECTS: FEDERAL AND STATE

The U.S. Food and Drug Administration has the responsibility of regulating animal feeds under the Food, Drug and Cosmetic Act of 1958, which in general specifies the same standards for human food and animal feeds. In 1967 the FDA published a Statement of Policy and Interpretation on the use of poultry litter in animal feed (Title 21 of the Code of Federal Regulations, Section 500.40). This policy statement outlined the FDA position regarding the nonsanctioning of this waste as a feed component on the grounds of possible drug residues and the possible transfer of disease organisms. Subsequently, the regulation was interpreted by the FDA to include other waste in addition to poultry litter. However, the agency also interpreted the regulations to mean that regulatory action would not be taken unless the animal waste intended for use in animal feed was found to be adulterated and was to be moved in interstate commerce. Since little animal feed is involved in interstate commerce, the FDA in practice left the responsibility for regulatory action to the states (Taylor and Geyer, 1979). In December 1980 the FDA published a document revoking its policy regarding feeding animal waste, leaving regulation of feeding animal waste to the states (Goyan, 1980).

The Association of American Feed Control Officials (AAFCO) has developed and revised a model feed bill and model regulations (Association of American Feed Control Officials, 1982:63). These regulations include "AAFCO Official Feed Terms and Feed Ingredient Definitions."

Some states have specific regulations providing for use of animal wastes as feed ingredients. These states include California, Colorado, Mississippi, Washington, Alabama, and Virginia. In addition, Georgia, Florida, Oregon, and Iowa have registered animal wastes as permitted feed ingredients. Two categories of animal wastes have been recognized under the state regulations, those with and those without drug residues. In some states a 15- or 30-day withdrawal period is required with the former.

The main legislation governing animal feeds in Canada is the Feeds Act (1967) which is currently under revision. The feeding of animal wastes is not allowed under this legislation (Jefferson, 1975) on the grounds of potential disease risk, possible mold and mycotoxin problems, and possible problems from residues. Related legislation governing animal products in Canada is the Food and Drugs Act, which specifies standards for drug and chemical residues in meat.

Dried layer waste has been given a tentative listing under the Canadian Feeds Regulations. It is defined as undiluted poultry excreta from layer flocks not receiving medication. The proposed clearance is for beef cattle and broiler (roaster) feeds at a level not exceeding 20 or 10 percent, respectively, with a withdrawal period of 15 days before slaughter.

RESEARCH NEEDS

Variability in composition is a notable feature of animal wastes, and research aimed at improved uniformity is required. This should include studies on collecting and processing systems and on feeding programs for the animals from which the waste is to be collected. Research aimed at improving and maintaining the nutritional value of animal wastes is also required.

An integrated animal systems approach should be taken in studies of utilization, for instance the housing of growing turkeys and beef animals

Animal Wastes 165

in adjacent lots. This would minimize handling and transport of wastes, and maximize nutrient utilization.

The high moisture content of animal waste needs to be reduced for most feeding systems, but dehydration may not be feasible in the future because of increased energy costs. Alternative ways of reducing dehydration requirements, such as predrying in the deep pits of poultry barns, need to be investigated. Research should also be conducted into the use of solar energy for dehydrators. This is distinct from the use of solar energy to produce sun-dried material, which has not had the benefit of pasteurization at a high temperature.

Ensiling is an inexpensive and effective method of processing wastes intended for animal feeding. Many organisms are killed by the process, but research needs to be carried out on methods aimed at preventing sporulation of spore-forming organisms.

Risks from residues do not appear to be a demonstrable problem in animals fed animal wastes, but research aimed at defining the appropriate withdrawal periods to ensure freedom from residues needs to be conducted. Medicinal drug residues and metabolites, mycotoxins, and mineral elements should be covered in this research. Further work also needs to be conducted on the potential for disease transmission by waste recycling, particularly botulism and related conditions.

Further economic studies need to be conducted on the various systems of feeding animal wastes that are currently or potentially in use to determine the overall economic benefits.

SUMMARY

Animal waste represents a feed resource that is presently not used to its nutritional and economic potential.

These wastes are of relatively low energy content, mainly due to a high content of fiber and ash. Consequently, they should be considered for inclusion in diets in which a high energy level is not of primary importance. Another feature of these wastes is that much of the nitrogen is normally present in nonprotein form. This feature, and the presence of fiber, indicates that animal wastes are more suited to recycling through ruminants, since these animals possess a digestive tract capable of effectively utilizing both fiber and nonprotein nitrogen. However, it has been shown that swine and poultry are capable of utilizing the true protein and other nutrients present in animal wastes. Animal wastes are variable in composition, and research aimed at improved uniformity and quality is required.

The wastes possessing the highest nutritive value are layer waste and broiler litter. They can be processed successfully by either dehydration or ensiling. These wastes contain as much true protein as the common feed grains and in addition are useful sources of calcium and phosphorus. The metabolizable energy value of poultry wastes for poultry is about one-half to one-third that of the common feed grains. There appears to be no problem of acceptability.

Animals best suited to utilize wastes are probably growing and finishing beef animals, beef breeding stock, growing dairy heifers, dry dairy cows, and sheep.

Wet wastes can be utilized successfully in silage systems and to some extent in liquid-feed systems.

Dehydrated wastes can be included successfully in high-dry-matter diets for a wide range of animals, and wastes processed in other ways can also be utilized successfully. For best results they should be formulated correctly into diets to prevent imbalances. Several workers have shown that the inclusion of too high a level of waste results in an excessive level of fiber and/or minerals in the diet, with a resultant depression in animal performance. Because of this limitation, up to 10 or 20 percent waste can be included successfully in some animal diets, such as high-energy diets. On the other hand, much higher levels can be used in diets for beef cows (80 percent). Productivity of animals fed diets containing animal wastes is high, and when the diets are fed correctly, animals demonstrate growth rates and production of milk, meat, and eggs equal to those of animals fed traditional feed ingredients.

There appears to be only a minimal disease risk with wastes that have been subjected to appropriate processing (dehydration or ensiling). One definable risk involves spore-forming bacteria, which are not destroyed by either process. Further research should be directed to this potential problem.

Copper appears to present the only definable mineral problem with the recycling of animal wastes, and only with respect to sheep. This is a quality control problem that requires action on the part of the feeder. The problem of excess copper in sheep diets is well known and is not connected solely with waste feeding.

Drug and chemical residues in tissues do not appear to present a major problem as a result of waste feeding, but more research should be conducted in this area. It is recommended that waste feeding should be followed by a withdrawal period of at least 15 days when the waste-fed animal is intended to provide milk, meat, or eggs for human consumption.

Food quality does not appear to be affected by feeding waste.

LITERATURE CITED

Alexander, D. C., J. A. J. Carriere, and K. A. McKay. 1968. Bacteriological studies of poultry litter fed to livestock. Can. Vet. J. 9:127.

Allred, J. N., J. W. Walker, V. C. Beal, and F. W. Germaine. 1967. A survey to determine the Salmonella contamination rate in livestock and poultry feeds. J. Am. Vet. Med. Assoc. 151:1857.

Andersen, H. R., M. Sorensen, J. Lykkeaa, and K. Kousgaard. 1976. Feeding Dried Poultry Waste for Intensive Beef Production. 443. Beret. Forsoegslab., Statens Husdrybrugsudvalg.

Anthony, W. B. 1966. Utilization of animal waste as feed for ruminants. Pp. 109–112 in Management of Farm Animal Wastes: Proceedings National Symposium. Publ. SP-0366. St. Joseph, Mich.: American Society of Agricultural Engineers.

Anthony, W. B. 1969. Cattle manure: Reuse through wastelage feeding. Pp. 293–296 in Livestock Waste Management and Pollution Abatement: Proceedings International Symposium on Livestock Wastes, Columbus, Ohio. St. Joseph, Mich.: American Society of Agricultural Engineers.

Anthony, W. B. 1970. Feeding value of cattle manure for cattle. J. Anim. Sci. 30:274.

Anthony, W. B. 1971. Cattle manure as feed for cattle. ASAE Pub. Proc. 271:293.

Anthony, W. B., and R. R. Nix. 1962. Feeding potential of reclaimed fecal residue. J. Dairy Sci. 45:1538.

Association of American Feed Control Officials. 1982. Model Regulation for Processed Animal Waste Products as Animal Feed Ingredients. Official Publication, Association of American Feed Control Officials. p. 63.

Batsman, V. 1973. Dried poultry droppings in feed for cattle. Molochnoe-Myasn. Skotovod. (Kiev) 6:28.

Berbeci, C., C. Rarinca, and D. Georgescu. 1975. Use of dried fowl droppings in the intensive fattening of lambs. Lucr. Stiint. Inst. Cercet. Nutr. Anim. 4:107.

Berryman, C. 1970. The problem of disposal of farm wastes with particular reference to maintaining soil fertility. P. 19 in Proceedings Symposium on Farm Wastes. Institute of Water Pollution Control, University of Newcastle-upon-Tyne.

Bevill, A. F., L. G. Biehl, M. Marshfield, and G. Koritz. 1978. Sulfonamide residues. Proceedings 27th Annual Texas A&M University Swine Shortcourse, April 3–5.

Bhargava, K. K., and J. B. O'Neil. 1975. Evaluation of dehydrated poultry waste from cage reared broilers as a feed ingredient for broilers. Poult. Sci. 54:1506.

Bhattacharya, A. N., and J. P. Fontenot. 1965. Utilization of different levels of poultry litter nitrogen by sheep. J. Anim. Sci. 24:1174.

Bhattacharya, A. N., and J. P. Fontenot. 1966. Protein and energy value of peanut hull and wood shaving poultry litters. J. Anim. Sci. 25:367.

Bhattacharya, A. N., and J. C. Taylor. 1975. Recycling animal waste as a feedstuff: A review. J. Anim. Sci. 41:1438.

Biely, J., R. Soong, L. Seier, and W. H. Pope. 1972. Dehydrated poultry waste in poultry rations. Poult. Sci. 51:1502.

Biely, J., and P. Stapleton. 1976. Recycled dried poultry waste in chick starter diets. Br. Poult. Sci. 17:5.

Bjornhog, G., and L. Sjoblom. 1977. Demonstration of coprophagy in some rodents. Swed. Agric. Res. 7:105.

Blair, R. 1974. Evaluation of dehydrated poultry waste as a feed ingredient for poultry. Fed. Proc. 33:1934.

Blair, R., and K. M. Herron. 1982. Growth performance of broilers fed diets containing processed poultry wastes. Br. Poult. Sci. 23:279.

Blair, R., and D. W. Knight. 1973. Recycling animal wastes. 1. The problems of disposal, and regulatory aspects of recycled wastes. 2. Feeding recycled wastes to poultry and livestock. Feedstuffs 45(10):32, 45(12):31.

Bohme, H. 1973. The possible use of dried poultry excreta in feeding. Landwirtsch. Forsch. 28:43.

Bohstedt, G., R. H. Grummer, and O. B. Ross. 1943. Cattle manure and other carriers of B-complex vitamins in rations for pigs. J. Anim. Sci. 2:373.

Borgioli, E., and M. Tocchini. 1969. Research on the use of sterilized poultry litter on beef-bullocks feeding. Aliment. Anim. 13:263.

Bosman, S. W. 1973. Chicken litter in fattening rations for cattle and sheep. S. Afr. J. Anim. Sci. 3:57.

Boubedja, M., and H. Marx. 1974. Studies on cattle fattening diets containing poultry litter. Food Sci. Technol. Abstr.—1975. 7 6S756.

Boyer, C. I., Jr., D. W. Bruner, and J. A. Brown. 1958. Salmonella organisms isolated from poultry feed. Avian Dis. 2:396.

Boyer, C. I., Jr., S. Narotsky, D. W. Bruner, and J. A. Brown. 1962. Salmonellosis in turkeys and chickens associated with contaminated feed. Avian Dis. 6:43.

Bressler, G. O., and E. L. Bergman. 1971. Solving the poultry manure problem economically through dehydration. ASAE Proc. 271:81.

Brugman, H. H., H. C. Dickey, B. E. Plummer, and J. Gooten. 1967. Drug residues in lamb carcasses fed poultry litter. J. Anim. Sci. 26:915. (Abstr.)

Bucholtz, H. F., H. E. Henderson, J. W. Thomas, and H. C. Zindel. 1971. Dried animal waste as a protein supplement for ruminants. ASAE Publ. PROC-271:308.

Bull, L. S., and J. T. Reid. 1971. Nutritive value of chicken manure for cattle. ASAE Proc. 271:297.

Caldertone, S. H., and H. A. Wilson. 1976. Some Microbial, Drying, and Odor Reduction Studies of Poultry Wastes. Bull. Agric. Exp. Stn., Univ. W. Va. No. 646T.

Calvert, C. C. 1973. Feed additive residues in animal manure processed for feed. Feedstuffs 45(17):32.

Calvert, C. C. 1975. Arsenicals in animal feeds and wastes. In Arsenical Pesticides. ACS Symp. Ser. No. 7:70.

Calvert, C. C. 1976. Systems for the indirect recycling by using animal and municipal wastes as a substrate for protein production. P. 245 in M. Chenost, ed. Proceedings of the Technical Consultation on New Feed Resources. Rome: FAO.

Calvert, C. C. 1979. Use of animal excreta for microbial and insect protein synthesis. J. Anim. Sci. 48:178.

Calvert, C. C., and L. W. Smith. 1972. Arsenic in milk and blood of cows fed organic arsenic compounds. J. Dairy Sci. 55:706. (Abstr.)

Calvert, C. C., and L. W. Smith. 1976. Heavy metal differences in tissues of dairy steers fed either cottonseed meal or dehydrated poultry excreta supplements. Proc. Annu. Meet., Am. Dairy Sci. Assoc. 127. (Abstr.)

Carriere, J. A. J., D. C. Alexander, and K. A. McKay. 1968. The possibility of producing tuberculin sensitivity by feeding poultry litter. Can. Vet. 9:178.

Caswell, L. F., J. P. Fontenot, and K. E. Webb, Jr. 1975. Effect of processing method on pasteurization and nitrogen components of broiler litter and on nitrogen utilization by sheep. J. Anim. Sci. 40:750.

Caswell, L. F., J. P. Fontenot, and K. E. Webb, Jr. 1978. Fermentation and utilization of broiler litter ensiled at different moisture levels. J. Anim. Sci. 46:547.

Cenni, B., G. Jannella, and B. Colombani. 1969. Poultry litter for feeding table poultry. Ann. Fac. Med. Vet. Pisa, Univ. Studi Pisa 22:276.

Chang, T. S., J. E. Dixon, M. L. Esmay, C. J. Flegal, J. B. Gerrish, C. C. Sheppard, and H. C. Zindel. 1975. Microbiological and chemical analyses of anaphage in a complete layer excreta in-house drying system. ASAE Proc. 275:206.

Chang, T. S., D. Dorn, and H. C. Zindel. 1974. Stability of poultry anaphage. Poult. Sci. 53:2221.

Ciordia, H., and W. B. Anthony. 1969. Viability of parasitic nematodes in wastelage. J. Anim. Sci. 28:133. (Abstr.)

Clark, J. L., M. R. Dethrow, and J. M. Vandepopuliere. 1975. Dried poultry waste as a supplemental nitrogen source for cattle. J. Anim. Sci. 41:394. (Abstr.)

Cohen, A., and R. Tamarin. 1978. Investigations of two mass outbreaks of a botulism-like disease in cattle. IV. Bacteriological investigations. Refu. Vet. 35:109.

Council for Agricultural Science and Technology. 1978. Feeding Animal Waste. Report No. 75. Ames: Iowa State University.

Cregar, C. R., F. A. Gardner, and F. M. Farr. 1973. Broiler litter silage for fattening beef animals. Feedstuffs 45:25.

Cross, D. L., and B. F. Jenny. 1976. Turkey litter silage in rations for dairy heifers. J. Dairy Sci. 59:919.

Cross, D. L., G. C. Skelley, C. S. Thompson, and B. F. Jenny. 1978. Efficacy of broiler litter silage for beef steers. J. Anim. Sci. 47:544.

Cuevas, S. 1969. Poultry droppings as a source of protein for fattening sheep. Rev. Mex. Prod. Anim. 2:27.

Cullison, A. E., H. C. McCampbell, A. C. Cunningham, R. S. Lowrey, E. P. Warren, B. D. McLendon, and D. H. Sherwood. 1976. Use of poultry manures in steer finishing rations. J. Anim. Sci. 42:219.

Day, D. L., and B. G. Harmon. 1974. A recycled feed source from aerobically processed swine wastes. Trans. ASAE 17:82.

Denisov, N. I., M. P. Kirilov, and N. A. Sorokin. 1973. Processed poultry droppings in feeds. Zhivotnovodstvo 2:45.

Denisov, N. I., M. P. Kirilov, Y. V. Abakumov, I. G. Tereshchenko, and I. Z. Reznikov. 1974. Poultry droppings as a component of the concentrate feed for pigs. Zhivotnovodstvo 6:54.

Denisov, N. I., M. P. Kirilov, L. A. Ilyukhina, V. Yu Abakumov, and A. K. Sabirov. 1975a. Supplements of poultry droppings in concentrates. Zhivotnovodstvo 12:43.

Denisov, N. I., M. P. Kirilov, Y. Abakumov, D. Popov, A. Slavin, and G. Rogov. 1975b. Dried poultry droppings in feeds for pigs. Svinovodstvo 5:18.

Diaz, C. P., and A. Elias. 1976. Cattle manure and final molasses silage in pig feeding. I. Effect of different levels in final molasses diets for growing pigs. Cuban J. Agric. Sci. 10:191.

Donoho, A. L. 1975. Metabolism of rumensin. In Proc., Rumensin Res. Seminar. Indianapolis: Eli Lilly.

Drake, C. L., W. H. McClure, and J. P. Fontenot. 1965. Effects of level and kind of broiler litter for fattening steers. J. Anim. Sci. 24:879 (Abstr.).

Dugan, G. L., C. G. Golueke, and W. J. Oswald. 1969. Hydraulic handling of poultry manure integrated into an algal recovery system. In Proceedings 1969 National Poultry Litter and Waste Management Seminar. 57 pp.

Dugan, G. L., C. G. Golueke, and W. J. Oswald. 1971. Poultry operation with an integrated sanitation waste materials recycling system. Abstracts, Excerpts and Reviews of the Solid Waste Literature, Vol. IV, prepared by C. G. Golueke. SERL Rep. 71-2:284.

Durham, R. M., G. W. Thomas, R. C. Albin, L. G. Howes, S. C. Curl, and T. W. Box. 1966. Coprophagy and use of animal waste in livestock feeds. ASAE Publ. SP-0366:112.

Egyed, M. N., A Shlosberg, U. Klopfer, T. A. Nobel, and E. Mayer. 1978a. Mass outbreaks of botulism in ruminants associated with ingestion of feed containing poultry waste. 1. Clinical and laboratory investigations. Refu. Vet. 35:93.

Egyed, M. N., U. Klopfer, T. A. Nobel, A. Shlosberg, A. Tadmor, I. Zukerman, and J. Avidar. 1978b. Mass outbreaks of botulism in ruminants associated with ingestion of feed containing poultry waste. II. Experimental investigation. Refu. Vet. 35:100.

Elmund, G. K., S. M. Morrison, D. W. Grant, and M. P. Nevins. 1971. Role of excreted chlortetracycline on modifying the decomposition process of feedlot waste. Bull. Environ. Contam. Toxicol. 6:129.

El-Sabban, F. F., J. W. Bratzler, T. A. Long, D. E. H. Frear, and R. F. Gentry. 1970. Value of processed poultry waste as a feed for ruminants. J. Anim. Sci. 31:107.

Esmay, M. L., C. J. Flegal, J. B. Gerrish, J. E. Dixon, C. C. Sheppard, H. C. Zindel, and T. S. Chang. 1975. Inhouse handling and dehydration of poultry manure from a caged layer operation: A project review. ASAE Publ. PROC-275:468.

Evans, E., E. T. Moran, Jr., and J. P. Walker. 1978a. Laying hen excreta as a ruminant feedstuff. I. Influence of practical extremes in diet, waste management procedures and stage of production on composition. J. Anim. Sci. 46:520.

Evans, E., E. T. Moran, Jr., G. K. Macleod, and E. M. Turner, Jr. 1978b. Laying hen excreta as a ruminant feedstuff. II. Preservation and acceptability of wet excreta by sheep. J. Anim. Sci. 46:527.

Evvard, J. M., and K. K. Henness. 1925. An experiment to study on hogs following cattle. P. 55 in Proc. Am. Soc. Anim. Prod.

Farquar, A. S., W. B. Anthony, and J. V. Ernst. 1979. Prevention of sporulation of bovine coccidia by the ensiling of a manure blended diet. J. Anim. Sci. 49:1331.

Faruga, A., H. Puchajda, and T. Mazur. 1974. Cattle manure in feeds for hens. Zesz. Nauk. Akad. Roln.-Techn. Olsztynie, Technol. Zywn. 129:79.

Feldhofer, S., E. Dumanovsky, M. Ostric, B. Rapic, D. Milosevic, B. Smalcelj, L. Milakovic-Novak, M. Lucic, A. Svalina, D. Haberstok, and A. Gjuric. 1975. Changes in poultry waste during processing and storage and its value in feeding ruminants. 2. Comparison of chemical analyses of fresh poultry waste kept for 7, 14 and 21 days. Stocarstvo 29:49.

Flachowsky, G. 1975. Studies in the suitability of solid materials in pig feces for use in the feeding of fattening cattle. 1. Procedures and results of fattening trials. Arch. Tierernaehr. 25:139.

Flachowsky, G. 1977. Incorporation of decanted solids of pig feces in feed for fattening cattle. 2. Comparison of different types of ration. Arch. Tierernaehr. 27:57.

Flachowsky, G., and H. J. Lohnert. 1974. Feeding value of the solids in rabbit feces. Arch. Tierernaehr. 24:611.

Flegal, C. J., and H. C. Zindel. 1971. Dehydrated poultry waste (DPW) as a feedstuff in poultry rations. Pp. 305–307 in Livestock Waste Management and Pollution Abatement: Proceedings International Symposium on Livestock Wastes, Columbus, Ohio. St. Joseph, Mich.: American Society of Agricultural Engineers.

Flipot, P., M. McNiven, and J. D. Summers. 1975. Poultry wastes as a feedstuff for sheep. Can. J. Anim. Sci. 55:291.

Fontenot, J. P., and V. Jurubescu. 1980. Processing of animal waste by feeding to ruminants. P. 641 in Proceedings of the 5th International Symposium on Ruminant Physiology. Lancaster, England: MTP Press.

Fontenot, J. P., and K. E. Webb. 1975. Health aspects of recycling animal wastes by feeding. J. Anim. Sci. 40:1267.

Fontenot, J. P., A. N. Bhattacharya, C. L. Drake, and W. H. McClure. 1966. Value of broiler litter as a feed for ruminants. ASAE Publ. SPO 366:105.

Fontenot, J. P., R. E. Tucker, B. W. Harmon, K. G. Libke, and W. E. Moore. 1970. Effects of feeding different levels of broiler litter to sheep. J. Anim. Sci. 30:319.

Fontenot, J. P., K. E. Webb, Jr., B. W. Harmon, R. E. Tucker, and W. E. C. Moore. 1971a. Studies of processing, nutritional value, and palatability of broiler litter for ruminants. Pp. 301–304 in Livestock Waste Management and Pollution Abatement: Proceedings International Symposium on Livestock Wastes. Columbus, Ohio. St. Joseph, Mich.: American Society of Agricultural Engineers.

Fontenot, J. P., K. E. Webb, Jr., K. G. Libke, and R. J. Bueler. 1971b. Performance and health of ewes fed broiler litter. J. Anim. Sci. 33-283. (Abstr.)

Fookes, R. F. 1972. The nutritive value of dried poultry manure for poultry. P. 77 in Proc. Australian Poultry Science Convention, Auckland. Auckland, N.Z.: World's Poultry Science Association.

Fuller, H. L. 1956. The value of poultry by-products as sources of protein and unidentified growth factors in broiler rations. Poult. Sci. 35:1143.

Galmez, de P. J., E. Santisteban, E. Haardt, C. Crempien, and L. Villalta. 1971a. Broiler chicken litter in feeds for fattening lambs. Agric. Tec. 31:24.

Galmez, de P. J., M. E. Santisteban, and D. R. Torell. 1971b. Broiler litter in feeds for gestating and lactating ewes. Agric. Tec. 31:208.

Geri, G. 1968. Growth, intake of feed and results of rearing of young pigs fed on mixtures containing poultry droppings. Aliment. Anim. 12:559.

Geri, G., M. Antongiovanni, and E. Sottini. 1970a. Effect of feeding young cattle on diets containing poultry litter on the fermentation characteristics of the microbial population of the rumen. Aliment. Anim. 14:27.

Geri, G., E. Sottini, and M. Antongiovanni. 1970b. Nutritive characteristics of poultry litter: Production of volatile fatty acids and utilization of constituents in the semi-permeable artifical rumen. Aliment. Anim. 14:21.

Geri, G., E. Sottini, and A. Olivetti. 1970c. Digestion in vitro of pure poultry droppings dried by different procedures. Aliment. Anim. 14:25.

Gilbertson, C. B., J. A. Nienaber, J. R. Ellis, T. M. McCalla, T. J. Klopfenstein, and S. D. Farlin. 1974. Nutrient and Energy Composition of Beef Cattle Feedlot Waste Fractions. Nebraska Agric. Exp. Stn. Res. Bull. 262.

Goering, H. K., and L. W. Smith. 1977. Composition of corn plant ensiled with excreta or nitrogen supplements and its effect on growing wethers. J. Anim. Sci. 44:452.

Goyan, J. F. 1980. Recycled animal waste. Fed. Reg. 45(251):86272.

Griel, L. C., Jr., D. C. Kradel, and E. W. Wickersham. 1969. Abortion in cattle associated with the feeding of poultry litter. Cornell Vet. 59:226.

Guedas, J. R. 1966. Basic investigations on the use of poultry droppings for feeding ruminants. Toxicity, digestibility, N balance and breakdown of uric acid. Rev. Nutr. Anim. Madrid. 4:11,128.

Guedas, J. R. 1967. Basic investigations on the use of poultry droppings for feeding ruminants. Toxicity, digestibility, N balance and breakdown of uric acid. Rev. Nutr. Anim. Madrid. 5:53,110.

Hamblin, D. C. 1980. Commercially processing and selling poultry waste as a feed ingredient. J. Anim. Sci. 50:342.

Harmon, B. G. 1974. Potential for recycling swine waste. Feedstuffs 46(9):40.

Harmon, B. G., D. L. Day, A. H. Jensen, and D. H. Baker. 1972. Nutritive value of aerobically sustained swine excrement. J. Anim. Sci. 34:403.

Harmon, B. G., D. L. Day, D. H. Baker, and A. H. Jensen. 1973. Nutritive value of aerobically and anaerobically processed swine waste. J. Anim. Sci. 37:510.

Harmon, B. W., J. P. Fontenot, and K. E. Webb, Jr. 1974. Effect of processing method of broiler litter on nitrogen utilization by lambs. J. Anim. Sci. 39:942.

Harmon, B. W., J. P. Fontenot, and K. E. Webb, Jr. 1975a. Ensiled broiler litter and corn forage. I. Fermentation characteristics. J. Anim. Sci. 40:144.

Harmon, B. W., J. P. Fontenot, and K. E. Webb, Jr. 1975b. Ensiled broiler litter and corn forage. II. Digestibility, nitrogen utilization and palatability by sheep. J. Anim. Sci. 40:156.

Harpster, H. W., T. A. Long, and L. L. Wilson. 1978. Comparative value of ensiled cattle waste for lambs and growing finishing cattle. J. Anim. Sci. 46:238.

Hauge, S., and K. Bovre. 1958. The occurrence of Salmonella bacteria in imported vegetable protein concentrates and mixed concentrates. Nord. Veterinaermed. 12:255.

Heichel, G. H. 1976. Agricultural production and energy resources. Am. Sci. 64:64.

Helmer, J. W. 1980. Monitoring the quality and safety of processed animal waste products sold commercially as feed. J. Anim. Sci. 50:349.

Hendrickson, D. A., and D. W. Grant. 1971. Aflatoxin formation in sterilized feedlot manure and fate during simulated water treatment practices. Bull. Environ. Contam. Toxicol. 6:525.

Hennig, A., D. Schuler, H. H. Freytag, C. Voigt, K. Gruhn, and H. Jeroch. 1972. Tests conducted to determine whether pig feces could be used as a feeding stuff. Jahrb. Tierernaehr. Futter. 8:226.

Hennig, A., H. Jeroch, H. J. Lohnert, and G. Flachowsky. 1975. Feed value for wethers of broiler and pullet excreta. Arch. Tierernaehr. 25:583.

Hesseltine, C. W. 1976. Conditions leading to mycotoxin contamination of foods and feeds. P. 1–22 in Mycotoxins and Other Fungal Related Food Problems, J. W. Rodricks, ed. Adv. Chem. Ser. 149. Washington, D.C.: American Chemical Society.

Hodgetts, B. 1971. The effects of including dried poultry waste in the feed of laying hens. Pp. 311–313 in Livestock Waste Management and Pollution Abatement: Proceedings International Symposium on Livestock Wastes, Columbus, Ohio. St. Joseph, Mich.: American Society of Agricultural Engineers.

Hsu, T. T. 1976. Wastelage as a feed for cattle. Wastelage made with rice husks, rice straw or sugar cane bagasse and cattle excreta. J. Taiwan Livestock Res. 9:135.

Hull, J. L., and J. B. Dobie. 1973. Feedlot animal waste compared with cottonseed meal as a supplement for pregnant rangecows. ASAE Paper No. 73-4506.

Jefferson, C. H. 1975. Regulating animal waste feeds. P. 219 in Waste Recycling and Canadian Agriculture. Ottawa: Agricultural Economics Research Council of Canada.

Jentsch, W., R. Schiemann, and H. Wittenburg. 1977. Feed energy value of pelleted fecal solids. Arch. Tieraenehr. 27:117.

Johnson, H. S., D. L. Day, C. S. Byerly, and S. Prawirokusumo. 1977. Recycling oxidation ditch mixed liquor to laying hens. Poult. Sci. 56:1339.

Johnson, R. R., R. Panciera, H. Jordan, and L. R. Shuyler. 1975. Nutritional, pathological and parasitological effects of feeding feedlot waste to beef cattle. ASAE Proc. 275:203.

Kanev, S., H. Krastanov, N. Nestorov, and T. Todorov. 1971. Comparative fattening of young cattle with complete feeds. Zhivotnuvd Nauki. 8:13.

Kazheka, V. I., and A. A. Kozyr. 1975. Dried poultry droppings (poudrette) in feeds for young sheep. Vestsi Akad. Navuk B. SSR, Ser. Sel'skagaspad. Navuk. 1:108.

Kienholz, E. W., G. M. Ward, J. M. Navallo, and M. C. Pritzl. 1975. Nutritional value of Cereco for poultry. Feedstuffs 47:21, 24.

Kneale, W. A., and J. R. Garstang. 1975. Milk production from a ration containing dried poultry waste. Exp. Husb. 28:18.

Knight, E. F., T. A. McCaskey, W. B. Anthony, and J. L. Walters. 1977. Microbial population changes and fermentation characteristics of ensiled bovine manure-blended rations. J. Dairy Sci. 60:416.

Koenig, S. E., E. E. Hatfield, and J. W. Spears. 1978. Animal performance and microbial adaptation of ruminants fed formaldehyde treated poultry waste. J. Anim. Sci. 46:490.

Kornegay, E. T., M. R. Holland, K. E. Webb, Jr., K. P. Bovard, and J. D. Hedges. 1977. Nutrient characterization of swine fecal waste and utilization of these nutrients by swine. J. Anim. Sci. 44:608.

Kraft, D. J., C. Olechowski-Gerhardt, J. Berkowitz, and M. S. Finstein. 1969. Salmonella in waste produced at commercial poultry farms. Appl. Microbiol. 18:703.

Kristensen, V. F., P. E. Andersen, G. K. Jensen, A. N. Fisker, and H. E. Birkkjaer. 1976. Feeding value of dried poultry waste for dairy cows. Faellesud valget fur Statens Mejeri-og Husdyr Brugsforsog. 2. Beret. Hillerod, Denmark.

Kubena, L. F., F. N. Reece, and J. D. May. 1973. Nutritive properties of broiler excreta as influenced by environmental temperature, collection interval, age of broilers and diet. Poult. Sci. 52:1700.

Lamm, D., L. E. Jones, D. C. Clanton, and J. K. Ward. 1976. Dehydrated poultry waste as a nitrogen source for cattle. J. Anim. Sci. 41:409. (Abstr.)

Lamm, W. D., K. E. Webb, Jr., and J. P. Fontenot. 1979. Ensiling characteristics, digestibility and feeding value of ensiled cattle waste and ground hay with and without sodium hydroxide. J. Anim. Sci. 48:104.

Lee, D. J. W., and R. Blair. 1973. Growth of broilers fed on diets containing dried poultry manure. Br. Poult. Sci. 14:379.

Lee, D. J. W., and W. Bolton. 1977. The laying performance of two strains of hens offered diets containing dried poultry manure during the laying stage. Br. Poult. Sci. 18:1.

Lee, D. J. W., R. Blair, and P. W. Teague. 1976. The effects on rearing and subsequent laying performance of rearer diets containing two levels of protein and dried poultry manure or urea. Br. Poult. Sci. 17:261.

Lee, P. K., and Y. F. Yang. 1975. Sun-dried chicken droppings as a feed for broilers. J. Taiwan Livestock Res. 8:27.

Lee, P. K., and Y. F. Yang. 1976. Sun-dried chicken droppings as a feed for laying Leghorn hens. J. Taiwan Livestock Res. 9:103.

Liebholz, J. 1969. Poultry manure and meat meal as a source of dietary nitrogen for sheep. Aust. J. Exp. Anim. Husb. 9:589.

Lipstein, B., and S. Bornstein. 1971. Value of dried cattle manure as a feedstuff for broiler chicks. Isr. J. Agric. Res. 21:163.

Littlefield, L. H., J. K. Bletner, and O. E. Goff. 1973. The effect of feeding laying hens various levels of cow manure on the pigmentation of egg yolks. Poult. Sci. 52:179.

Lovett, J. 1972. Toxigenic fungi from poultry feed and litter. Poult. Sci. 51:309.

Lovett, J., J. W. Messer, and R. B. Read, Jr. 1971. The microflora of Southern Ohio poultry litter. Poult. Sci. 50:746.

Lowman, B. G., and D. W. Knight. 1970. A note on the apparent digestibility of energy and protein in dried poultry excreta. Anim. Prod. 12:525.

Lucas, D. M., J. P. Fontenot, and K. E. Webb, Jr. 1975. Composition and digestibility of cattle fecal waste. J. Anim. Sci. 41:1480.

Madsen, H. 1939. Does the rabbit chew the cud? Nature (London) 143:981.

Masters, G. C. 1977. Preservation and refeeding poultry waste to cattle. Proceedings Seminar on Feedlot Manure Recycling for Nutrient Recovery, U.S. Environmental Protection Agency, Ada, Okla. PGM 1–3, April 7.

McCaskey, T. A., and W. B. Anthony. 1975. Health aspects of feeding animal waste conserved in silage. ASAE Proc. 275:230.

McCaskey, T. A., and W. B. Anthony. 1979. Human and animal health aspects of feeding livestock excreta. J. Anim. Sci. 48:163.

McClarin, R., K. W. Newell, and C. R. Murdock. 1959. Salmonellosis in Northern Ireland with special reference to pigs and Salmonella contaminated pig meal. J. Hyg. 57:92.

McClure, W. H., J. P. Fontenot, and K. E. Webb, Jr. 1979. Ensiled corn forage and broiler litter and zeronal implant for finishing heifers. Va. Polytech. Inst. State Univ. Res. Div. Rep. 175:69.

McNab, J. M., D. W. F. Shannon, and R. Blair. 1974. The nutritive value of a sample of dried poultry manure for the laying hen. Br. Poult. Sci. 15:159.

McNiven, M., J. D. Summers, and S. Leeson. 1976. Liquid diets containing poultry wastes for ruminants. Can. J. Anim. Sci. 56:221.

Mello, R. D. de, F. E. Galvao, J. A. de F. Veloso, and R. F. Barbosa. 1973. Efficiency of chicken litter compared to cottonseed meal as source of protein of lactating cows. Arq. Esc. Vet. Univ. Fed. Minas Gerais. 25:143.

Meregalli, A., A. Olivetti, M. Antongiovanni, E. Sottini, and M. Aleandri. 1971. The use of dried poultry droppings in producing light young bulls. Aliment. Anim. 15:37.

Meregalli, A., E. Sottini, and A. Olivetti. 1973. Feeding young cattle on maize silage at the wax stage. Inclusion of poultry litter. Riv. Zootec. Vet. 1:57.

Merwe, H. J. Van Der, P. S. Pretorius, and J. E. J. Du Toit. 1975. Use of poultry droppings in growing rations for lambs. Agroanimalia 7:65.

Messer, J. W., J. Lovett, G. K. Murphy, A. J. Wehby, M. L. Schafer, and R. B. Read, Jr. 1971. An assessment of some public health problems resulting from feeding poultry litter to animals. Microbiological and chemical parameters. Poult. Sci. 50:874.

Miner, J. R., L. Boersma, J. E. Oldfield, and H. K. Phinney. 1975. Swine waste nutrient recovery system based on the use of thermal discharges. ASAE Proc. 275:160.

Muftic, R., D. Bugarski, M. Varadin, and M. Dzinic. 1974. Effect of broiler litter as basic component of the ration on productivity and reproduction of cows. Veterinaria (Yugoslavia) 23:397.

Muftic, R., M. Dzinic, D. Bugarski, and L. Matekalo. 1968. Nutritive value of poultry litter in feeding ruminants. 2. Vet. Glas. 22:929.

Muller, Z. O. 1976. Economic aspects of recycled wastes. P. 265 in M. Chenost (ed.) Proceedings of Technology Consulting on New Feed Resources. Rome: FAO.

Newton, G. L., P. R. Utley, R. J. Ritter, and W. C. McCormick. 1977. Performance of beef cattle fed wastelage and digestibility of wastelage and dried waste diets. J. Anim. Sci. 44:447.

Noland, P. R., B. F. Ford, and M. L. Ray. 1955. The use of ground chicken litter as a source of nitrogen for gestating-lactating ewes and fattening steers. J. Anim. Sci. 14:860.

Ochoa, M. A., F. O. Bravo, and R. A. Carrillo. 1972. Organic residues in the feeding of growing sheep. Tecnica Pecuaria en Mexico 22:11.

Oliphant, J. M. 1974. Feeding dried poultry waste for intensive beef production. Anim. Prod. 18:211.

Oltjen, R. R., and D. A. Dinius. 1976. Processed poultry waste compared with uric acid, sodium urate, urea, and biuret as nitrogen supplements for beef cattle fed forage diets. J. Anim. Sci. 43:201.

Orr, D. E., E. R. Miller, P. K. Ku, W. G. Bergen, D. E. Ullrey, and E. C. Miller. 1973. Swine waste as a nutrient source for finishing pigs. Mich. State Univ. Rep. Swine Res. 232:AH-SW-7319:81.

Osborne, T. B., and L. B. Mendel. 1914. The contribution of bacteria to the feces after feeding diets free from indigestible components. J. Biol. Chem. 18:177.

Osterc, J. 1972. Possibility of using dried poultry manure for feeding pigs. Zb. Bioteh. Fak. Univ. Ljubljani 19:175.

Ostrander, C. E. 1975. Techniques that are solving pollution problems for poultrymen. ASAE Proc. 275:71.

Parigi-Bini, R. 1969. Utilization of pure dried poultry litter (Toplan) in feeding sheep. Aliment. Anim. 13:277.

Pearce, G. R. 1975. The inclusion of pig manure in ruminant diets. ASAE Proc. 275:218.
Pequegnat, C. 1975. Economic feasibility of waste as animal feed. P. 26 in Waste Recycling and Canadian Agriculture. Ottawa: Agricultural Economics Research Council of Canada.
Pereira, W. M., J. C. A. de Mattos, C. Barbosa, A. C. M. de F. Siqueira, L. R. M. da Silva, and C. do A. Cintra. 1972. Performance and carcass yield of young Swiss-Gujerat bulls fattened in confinement on a ration based on manure from laying hens, dried in the shade. Bol. Indust. Anim. 29:1.
Perez-Aleman, S., D. G. Dempster, P. R. English, and J. H. Topps. 1971. A note on dried poultry manure in the diet of the growing pig. Anim. Prod. 13:361.
Polidori, F., G. D'Urso, and A. Lanza. 1972. Chemical and microbiological characteristics of dried poultry droppings and their use in diets for laying hens. Avicoltura 41:85.
Pomeroy, B. S., and M. K. Grady. 1960. The isolation of Salmonella organisms from feedstuffs. P. 158 in Proceedings of the 3rd National Symposium on Nitrofurans in Agriculture. Lexington: University of Kentucky.
Post, G., and G. M. Ward. 1975. Use of Cereco II in rainbow trout rations. Feedstuffs 47:24.
Pryor, W. V., and J. K. Connor. 1964. A note on the utilization by chicken of energy from feces. Poult. Sci. 43:833.
Rhodes, D. N. 1972. Eating quality of beef from steers fed poultry waste. Memorandum, Meat Research Institute. No. 12. Langford, Bristol, U.K.
Richter, M. F., R. L. Shirley, and A. Z. Palmer. 1980. Effect of roughage fraction of cattle manure on digestibility and net energy of feedlot diets fed steers. J. Anim. Sci. 50:207.
Saedi, H., and M. A. Zohari. 1968. Dried cattle manure in feeds for laying hens. J. Vet. Fac. Univ. Teheran 24:42.
Salo, M. L., M. Suvitie, and M. Nasi. 1975. Energy, protein and mineral value of dried poultry manure for sheep. J. Sci. Agric. Soc. Finl. 47:462.
Saylor, W. W., and T. A. Long. 1974. Laboratory evaluation of ensiled poultry waste. J. Anim. Sci. 39:139. (Abstr.)
Schake, L. M., B. W. Pinkerton, C. E. Donnell, J. K. Riggs, and R. E. Lichtenwalner. 1977. Utilization of cattle excrement for growth and maintenance of beef cattle. J. Anim. Sci. 45:166.
Shannon, D. W. F., and W. O. Brown. 1969. Losses of energy and nitrogen on drying poultry excreta. Poult. Sci. 48:41.
Shannon, D. W. F., R. Blair, and D. J. W. Lee. 1973. Chemical and bacteriological composition and the metabolisable energy value of eight samples of dried poultry waste produced in the United Kingdom. In Proceedings of 4th European Poultry Conference. London. 487 pp.
Sheppard, C. C., C. J. Flegal, H. C. Zindel, T. S. Chang, J. B. Gerrish, M. L. Esmay, and F. Walton. 1975. Modifications of the Michigan State poultry in-house drying system. ASAE Proc. 275:74.
Silva, L. A., H. H. Van Horn, E. A. Olaloku, C. J. Wilcox, and B. Harris, Jr. 1976. Complete rations for dairy cattle. VII. Dried poultry waste for lactating cows. J. Dairy Sci. 59:2071.
Singh, S. P. 1974. Microbiology of common feed ingredients used by the poultry industry. Feedstuffs 46(25):30.
Smith, L. W. 1973. Recycling animal wastes as protein sources. P. 146 in Alternative Sources of Protein for Animal Production. Washington, D.C.: National Academy of Sciences.
Smith, L. W., and C. C. Calvert. 1976. Dehydrated broiler excreta versus soybean meal as nitrogen supplements for sheep. J. Anim. Sci. 43:1286.

Smith, L. W., and I. L. Lindahl. 1977. Alfalfa versus poultry excreta as nitrogen supplements for lambs. J. Anim. Sci. 44:152.

Smith, L. W., and W. E. Wheeler. 1979. Nutritional and economic value of animal excreta. J. Anim. Sci. 48:144.

Smith, L. W., G. F. Fries, and B. T. Weinland. 1976. Poultry excreta containing polychlorinated biphenyls as a protein supplement for lactating cows. J. Dairy Sci. 59:465.

Smith, L. W., C. C. Calvert, and H. R. Cross. 1979. Dehydrated poultry excreta vs. cottonseed meal as nitrogen supplements for Holstein steers. J. Anim. Sci. 48:633.

Smith, O. B., G. K. Macleod, D. N. Mowat, C. A. Fox, and E. T. Moran, Jr. 1978. Performance and health of calves fed wet caged layer excreta as a protein supplement. J. Anim. Sci. 47:833.

Smith, O. B., G. K. Macleod, D. N. Mowat, and E. T. Moran, Jr. 1979. Effect of feeding organic acid treated hen excreta upon performance, carcass merit and health of feedlot cattle. J. Anim. Sci. 49:1183.

Sommer, A., and O. Pelech. 1971. Untreated broiler litter in feeds for fattening cattle. Pol'nohospodarstvo 17:584.

Southwell, B. L., O. M. Hale, and W. C. McCormick. 1958. Poultry house litter as a protein supplement in steer fattening rations. Ga. Agric. Exp. Stn. Mimeo Ser. 55.

Stapleton, P., and J. Biely. 1975. Utilization of dried poultry waste in chick starter rations. Can. J. Anim. Sci. 55:595.

Surbrook, T. C., C. C. Sheppard, J. S. Boyd, H. C. Zindel, and C. J. Flegal. 1971. Drying poultry waste. ASAE Proc. 271:192.

Suttle, N. F., and J. Price. 1976. The potential toxicity of copper-rich animal excreta to sheep. Anim. Prod. 23:233.

Swingle, R. S., A. Araiza, and A. R. Urias. 1977. Nitrogen utilization by lambs fed wheat straw alone or with supplements containing dried poultry waste, cottonseed meal or urea. J. Anim. Sci. 45:1435.

Szelenyi, E., E. Barabas, J. Czako, and J. Regius. 1971. Making use of poultry litter in fattening cattle. Scientific works of the Research Institute for Animal Production at Nitra 9:119.

Taiganides, E. P., and R. L. Stroshine. 1971. Impact of farm animal production and processing on the total environment. Pp. 95–98 in Livestock Waste Management and Pollution Abatement: Proceedings International Symposium on Livestock Wastes, Columbus, Ohio. St. Joseph, Mich.: American Society of Agricultural Engineers.

Taylor, J. C., and R. E. Geyer. 1979. Regulatory considerations in the use of animal waste as feed ingredients. J. Anim. Sci. 48:218.

Thomas, J. W., Y. Yu, P. Tinnimit, and H. C. Zindel. 1972. Dehydrated poultry waste as feed for milking cows and growing sheep. J. Dairy Sci. 55:1261.

Thorlacius, S. O. 1976. Nutritional evaluation of dehydrated cattle manure using sheep. Can. J. Anim. Sci. 56:227.

Tinnimit, P., Y. Yu, K. McGuffrey, and J. W. Thomas. 1972. Dried animal waste as a protein supplement for sheep. J. Anim. Sci. 35:431.

Trakulchang, N., and S. L. Balloun. 1975a. Effects of recycling dried poultry waste on young chicks. Poult. Sci. 54:615.

Trakulchang, N., and S. L. Balloun. 1975b. Use of dried poultry waste in diets for chickens. Poult. Sci. 54:609.

Umstadter, L. W. 1980. A unique system for nutrient utilization of cattle waste. J. Anim. Sci. 50:345.

Van Dyne, D. C., and C. B. Gilbertson. 1978. Estimating U.S. livestock and poultry manure and nutrient production. ESCS-12. Washington, D.C.: U.S. Department of Agriculture.

Van Soest, P. J., and J. B. Robertson. 1976. Composition and nutritive value of uncommon feedstuffs. In Proceedings Cornell Nutrition Conference. 102 pp.

Velloso, L., E. Roverso, B. C. Alves, and F. L. Lopes. 1970. Chicken litter as a source of protein for fattening cattle in confinement. Bol. Indust. Anim. 27:337.

Vetter, D. L. 1972. Role of recycled large animal wastes for animal protein production. P. 235 in Proceedings Iowa State University Nutrition Symposium on Protein. Ames: Iowa State University.

Vijchulata, P., P. R. Henry, C. B. Ammerman, H. N. Becker, and A. Z. Palmer. 1980. Performance and tissue mineral composition of ruminants fed cage layer manure in combination with monensin. J. Anim. Sci. 50:48.

Vogt, H. 1973. Dried poultry droppings in feeds for laying hens. Arch. Geflüegelkd. 37:141.

Wadleigh, C. H. 1968. Wastes in relation of agriculture and forestry. USDA Misc. Pub. 1065. Washington, D.C.: U.S. Department of Agriculture.

Ward, G. M., D. E. Johnson, and R. D. Boyd. 1975. Nutritional value of Cereco silage and Cereco protein for steers. Feedstuffs 47:19.

Webb, K. E., Jr., and J. P. Fontenot. 1975. Medicinal drug residues in broiler litter and tissue from cattle fed litter. J. Anim. Sci. 41:1212.

Webb, K. E., Jr., J. P. Fontenot, and W. H. McClure. 1979. Performance and liver copper levels of beef cows fed broiler litter. P. 109 in Livestock Research Report, Research Division Report 175. Blacksburg: Virginia Polytechnic Institute and State University.

Westing, T. W., and B. Brandenberg. 1974. Beef feedlot waste in rations for beef cattle. P. 336 in Proceedings of the Conference on Processing and Management of Agricultural Wastes. Ithaca, N.Y.: Cornell University Press.

Westing, T. W., J. P. Fontenot, K. E. Webb, Jr., and W. H. McClure. 1977. Mineral profiles in broiler litter and in liver and loin from finishing heifers fed ensiled broiler litter and corn forage. P. 101 in Research Division Report 172. Blacksburg: Virginia Polytechnic Institute and State University.

Will, L. A., S. L. Diesch, and B. S. Pomeroy. 1973. Survival of *Salmonella typhimurium* in animal manure disposal in a model oxidation ditch. Am. J. Pub. Health 63:322.

Yoshida, M., and H. Hoshii. 1968. Nutritional value of poultry manure. Jpn. Poult. Sci. 5:37.

Zgajnar, J. 1975. Digestibility of feeds of maize and poultry droppings. Wirtschaftseigene. Futter 21:233.

5
Crop Residues

INTRODUCTION

The potential for animal production through feeding crop residues is impressive. The United States produces annually about 44 million tons of wheat, 45 million tons of soybeans, and another 186 million tons of coarse feed grains, dry basis (U.S. Department of Agriculture, 1979). The feed grains consist primarily of corn, with lesser amounts of milo, barley, and oats. Grain-producing plants normally produce an equal or greater weight as vegetative material than as grain (Vetter and Boehlje, 1978). Therefore, a total of at least 300 million tons of straws, stalks, and stubbles are available in the United States each year (see Table 48) and another 40 million in Canada. Worldwide, over 1.5 billion tons of crop residues are produced each year (for definitions of residues and other terms see the glossary on p. 210).

The 300 million tons of residues have sufficient energy to meet the entire needs of the present beef cattle industry in the United States. However, the concentration of digestible energy is sufficiently low to prevent the practical use of much of the residue without further treatment or processing.

QUANTITY

Corn is the most widely produced grain crop in the United States. Normally, an amount of residue greater than the quantity of grain is produced above ground. Therefore, the production of 153 million tons of corn per year yields at least 153 million tons of corn residue, which is over one-half the total available residue supply (see Table 48).

Wheat residue constitutes 15 percent of the total crop residue because grain production per land area is generally lower for wheat. Over 44 million tons of wheat straw are produced per year.

Soybean residue accounts for another 15 percent of the crop residue (45 million metric tons) and grain sorghum 5 percent (16 million metric tons). Other small grains and crops account for the remaining residues (see Table 48).

Crop residues are widely distributed. Corn residue is distributed throughout the Corn Belt. Illinois, Iowa, Indiana, Kansas, Minnesota, and Nebraska produce about 50 percent of the total crop residue supply. Wheat straw is more widely distributed over the United States, but tends to be concentrated in the Great Plains States and the Northwest.

Soybean production has been centered in the Corn Belt, but is expanding in the Southeast. Grain sorghum is produced primarily in the Great Plains States, with 77 percent produced in Texas, Kansas, and Nebraska. Peanuts, cotton, and rice are grown in the southern United States and California, and sugarcane is produced in Louisiana, Florida, Hawaii, and Texas.

PHYSICAL CHARACTERISTICS

Crop residues can be collected from the field, but because of their bulky characteristics and low market value per unit weight, transportation is generally uneconomical. Because of differences in plant structure, grain harvesting methods, and moisture content, residues vary concerning ease of harvest. Straws from small grains are easily collected dry behind the combine. Corn and especially grain sorghum residues often do not dry to the extent needed for dry storage, but they can be stored as silage. Soybean residue is difficult to harvest if allowed to drop on the ground behind the combine. Soil contamination during harvest can be a problem with all residues. Feeding of residues to livestock on the farm can be readily accomplished. Most of the residues are produced in the summer and fall; storage for extended periods may not be economical, but the greatest use for livestock feed generally occurs in the winter.

Removal of crop residues from the soil may reduce soil tilth (organic matter) and increase the risk of wind and water erosion. Depending upon soil type and topography, an average of one-half the total residue can be removed (Larson et al., 1978). If the residue is used on the farm and the resulting manure returned to the soil or if animals consume the residues by grazing, there should be little detrimental effect on soil from crop residue feeding.

TABLE 48 Estimated Supply of Crop Residues

Crop Source	Dry Grain (Product) Production (tons × 10⁶)[a]	Residue Coefficient[b]	Dry Residue (tons × 10⁶)	Percent of Total
Corn	153.0	1	153.0	51.0
Soybean	45.2	1	45.2	15.0
Wheat	44.2	1	44.2	14.7
Grain Sorghum	16.2	1	16.2	5.4
Barley	8.8	2	17.6	5.9
Oats	7.9	1	7.9	2.6
Cotton	2.1	3	6.3	2.1
Rice	5.6	1	5.6	1.9
Peanuts	1.6	1.5	2.4	0.8
Flax	0.3	3	0.9	0.3
Rye	0.6	1	0.6	0.2
Sugar Beets	4.3	0.14	0.6	0.2
Total			300.5	

[a] From 1978 USDA Agricultural Statistics (1979).
[b] From Vetter and Boehlje (1978); cotton, peanuts, sugar beets, and flax from Corkern et al. (1979).

Some of the residues, such as cotton by-products, rice milling by-products, and bagasse, that are produced at central locations have the advantage of being collected and available for processing or treatment. In fact, they may initially have a negative value because they present a disposal problem.

NUTRITIVE VALUE

Without treatment, crop residues are low in nutritional value. Because grain is harvested after the plant reaches physiological maturity, the vegetative portion is high in cell walls and lignin and low in protein and digestible dry matter. It has been clearly shown that lignin inhibits digestion of cellulose and hemicellulose, making about 1.4 times the weight of the lignin in the matter indigestible (Van Soest, 1981). There is considerable variation in digestibility both among and within residues. Controlling and properly accounting for this variation may be the most important key to efficient use of crop residues.

Corn

Corn residue consists of 54 percent stalks, 12 percent leaves, 21 percent cobs, and 13 percent husks (Vetter, 1973). Husks and cobs are discharged from the rear of the combine and the mixture is referred to as husklage. The husk is the most digestible part of the corn plant, with the digestibility of the other parts being lower and more variable. Dry-matter digestibility of the various parts of the corn plant ranges from 40 to 70 percent and protein content from 2 to 7.5 percent (see Appendix Table 1). The corn residues contain 70 to 80 percent cell walls, with nearly equal quantities of cellulose and hemicellulose. Corn residue tends to be one of the richest sources of hemicellulose found in nature.

The cornstalk decreases in digestibility and cell-soluble content with time after physiological maturity of the corn plant (Berger et al., 1979; McDonnell and Klopfenstein, 1980) (see Figure 11).

Digestibility of cornstalks is affected by moisture, variety, and temperature, as well as by maturity (McDonnell and Klopfenstein, 1980). Stalks left in the field for grazing also decrease in nutritional value with time (Lamm and Ward, 1977). There is the possibility that the corn plant might be changed genetically to improve stalk digestibility. The brown midrib mutant is an example of such potential, but grain yields have been lower than for normal corn (Colenbrander et al., 1973; Karn and Hendrix, 1977).

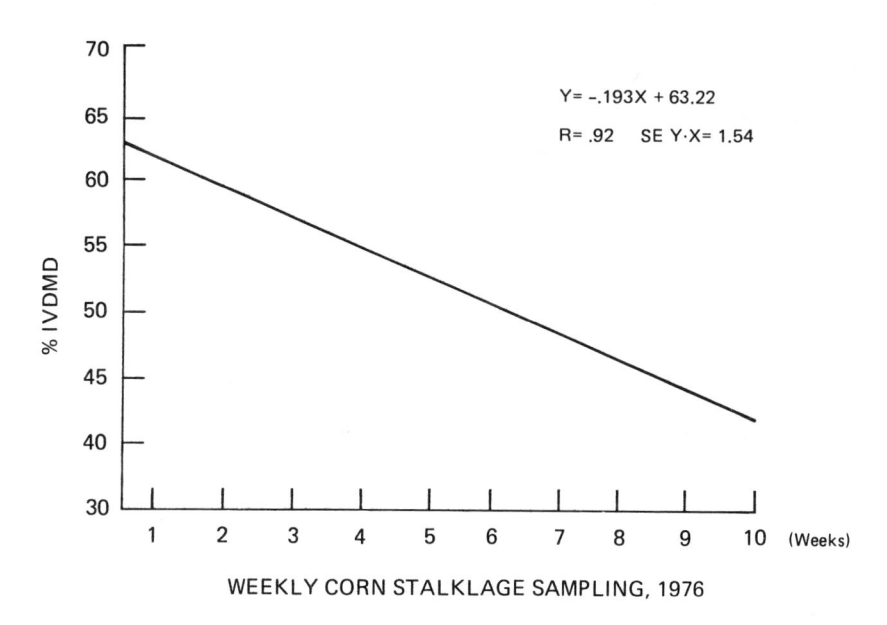

FIGURE 11 In vitro dry-matter disappearance (IVDMD) of cornstalks harvested over time. SOURCE: Berger et al. (1979).

Different harvesting systems produce corncobs, husklage, and cornstalks and stalklage; and stalk fields are available for grazing. Gestating beef cows are able to more than maintain weight on these residues (see Table 49). Growing calves gained weight (0.3 kg/day) on supplemented corncobs (Koers et al., 1970). Berger et al. (1979) obtained gains of 0.65 kg/day with growing calves fed supplemented stalklage harvested the same day as high-moisture (26 percent) grain. However, digestibilities as low as 36.9 percent have been reported (Paterson et al., 1979), which would not support maintenance in calves.

Wheat and Other Small Grain and Grass Straws

Small grain straws are quite low in nutritional value; in addition to being high in lignin content and, therefore, low in digestibility, they are also very bulky. This tends to limit intake and adds to mechanical handling problems. Wheat straw is probably the poorest quality straw, and barley is only slightly better (see Table 49). Oat straw appears to have the highest nutritional value (Anderson, 1978).

Acock (1978) has shown considerable variation in nutritional value of wheat straw (protein and digestibility). Location of production, variety, and time of straw harvest relative to grain harvest were all factors affecting value. Jackson (1978) also reported these variations, as well as that due to cultural practice.

Wheat straw with only a protein-vitamin-mineral supplement will not maintain weight of gestating beef cows or calves. Acock et al. (1979) and Dinusson (1969) have shown that feeding one-third alfalfa hay with wheat straw can meet the protein needs of the gestating beef cow and give some weight gain (0.2 kg/day). Coombe et al. (1979a) and Lesoing et al. (1980a) have clearly demonstrated that chopped untreated wheat straw has little value for growing calves. Similar conclusions can be drawn for barley straw. Oat straw can probably meet the energy needs of the gestating beef

TABLE 49 Dry Cow Daily Weight Gain (kg) on Various Corn Residue Systems (number of trials in parentheses)

Source	Grazing	Stacked Stalks	Stalklage	Shucklage
Iowa	0.03 (5)	0.40 (4)	0.08 (1)	0.20 (3)
Nebraska	0.25 (6)	0.26 (2)	0.31 (1)	—

SOURCE: Ward (1978).

cow but has little value for growing animals (Saxena et al., 1971) unless treated to improve its nutritional value.

Rice straw is somewhat unique in that lignin and high silicon content tend to limit digestibility. However, its feeding value would be similar to wheat straw (Garrett et al., 1979). Grass straws available after grass seed harvest have variable nutritive values. The variation is due to species (Guggolz et al., 1971), variety (Early and Anderson, 1978), and time of harvest (Durham and Hinman, 1979). Bluegrass straw supported small weight gains and normal production in gestating beef cows (Early and Anderson, 1978), but considerable amounts of supplemented grain were needed to produce 0.5 kg/day gain in calves (Durham and Hinman, 1979). Rice and grass seed straws must often be removed from the field for agronomic reasons. Therefore, these straws present disposal problems for producers.

Soybean

Soybean residue is 30 to 35 percent pods and the remainder stalks (Vetter, 1973). Essentially no leaves are collectable. The pods are lower in cell walls and lignin content than stalks and higher in protein and dry-matter digestibility. Soybean stalks are very high in lignin content (20 to 23 percent) and therefore very low in digestibility. The pods may have some use in beef cow maintenance diets, but probably have little value for growing or lactating ruminants (Gupta et al., 1978).

Grain Sorghum

The grain sorghum plant is unique in that it does not die at physiological maturity but continues to photosynthesize until a killing frost (Smith, 1977). The residue following grain harvest is green and moist (60 to 70 percent H_2O), and quality changes little with time (see Appendix Table 1). The moist residue harvested and stored as silage is a good beef cow feed (McKee et al., 1977) and has potential in a diet for growing calves. Both protein content and dry-matter digestibility may be higher than for any residue except early harvested cornstalks or cornhusks.

Bolsen et al. (1977) obtained over 0.5 kg/day gain with growing calves fed milo stover silage with protein supplement. They found an interesting and consistent positive associative effect of feeding this residue with forage sorghum silage (Bolsen et al., 1976). Ward et al. (1979) found that beef cows could be maintained during gestation on milo stubble fields. The cows gained about 0.3 kg/day, which is similar to the gain obtained with corn stalk grazing.

Other Residues

In the ginning of cotton, 50 to 75 percent of the harvested material is residue, consisting of boll residues, leaves, stems, lint, and a small quantity of cottonseeds. The residue has a low nutritional value, characterized by its high lignin (14 percent) and low protein (5.6 percent) content. Use of gin trash with supplemental cottonseed meal for growing steers produced gains of 0.1 kg/day compared with 0.5 kg/day for steers fed sorghum silage (Holloway et al., 1974). Finishing steers gained more rapidly when the roughage portion of the diet consisted of 50 percent cottonseed hulls than when it contained 50 percent cotton gin trash (Jones et al., 1957). The estimated value of gin trash was 90 percent that of cottonseed hulls. In another study, gin trash was substituted for 10, 30, and 50 percent of alfalfa hay cubes in a diet consisting of 50 percent alfalfa and 50 percent concentrate (Brown et al., 1979). Milk production of dairy cattle was depressed slightly at the two higher levels. Efficiency of converting feed energy to milk energy was decreased with increasing levels of gin trash.

Cottonseed hulls are a high-fiber, low-protein by-product of the cotton industry. They have been used as a roughage source for beef and dairy cattle. Cottonseed hulls substituted for alfalfa hay cubes at 10, 30, or 50 percent of the alfalfa resulted in no difference in milk production of dairy cattle (Brown et al., 1977). However, as the percentage of cottonseed hulls in the diet increased, the digestibilities of protein, energy, and fiber decreased. Hunt et al. (1971) also observed no differences in milk production when cottonseed hulls were included at 25, 35, and 45 percent of the diet. A comparison of bluegrass straw, wheat screenings, and pelleted and nonpelleted cottonseed hulls as roughage sources for yearling steers showed cottonseed hulls to be inferior to the other roughage sources (Heinemann, 1976). When each of the four sources constituted 13 percent of the diet, average daily gains were 1.4, 1.4, 1.3 and 1.2 kg/day, respectively. Feed conversion was least efficient for animals fed the cottonseed hull diets.

Although high in fiber and lignin content, peanut hulls have a considerable amount of protein (8 percent). This protein value may vary with the quantity of peanut kernels that remain with the hulls during processing. Utley and McCormick (1972) examined the use of peanut hulls in a feedlot study with Hereford steers. The hulls were combined with concentrate at 0, 10, 20, and 30 percent of the diet and fed ad libitum for 84 days. The highest average daily gains, when adjusted for body fill, were on diets with 10 or 20 percent hulls, and the most efficient feed conversion was with the 10 percent hull diet. A second feedlot study showed no significant differences among diets. Dry-matter digestibility decreased with the addition of peanut hulls. Calhoun and Shelton (1973) also found that 10 percent peanut hulls was optimum for high-concentrate lamb diets. The use of 30 percent peanut hulls for growing steers was found equivalent to 30 percent cottonseed hulls (Burdick et al., 1975), and growth of wintering calves was satisfactory with a 60 percent peanut hull diet.

Rice hulls are low in protein, high in fiber, and contain unusually large amounts of lignin (11–17 percent) and silica (22 percent). At 40 percent of the diet, scouring, with mucous and blood in the feces, has been observed with cattle (White, 1966). At 20 percent of the diet these symptoms were not seen. McManus and Choung (1976) reported that raw rice hulls were rejected by sheep. Wintering steers accepted rice hulls at 25 percent of the diet, the remainder being prairie hay (Noland and Ford, 1954). Daily feed intake was restricted to 5.4 kg, and average daily gain was 0.8 kg, compared with 0.86 for an all prairie hay diet.

Sugarcane bagasse is the residue of the crushing and milling of sugarcane. It is very low in protein (<2 percent) and high in fiber. Randel (1970a) was unable to show any differences in performance of male calves initially weighing 114 to 454 kg fed a 20 or 30 percent bagasse diet at either 12.5 or 16 percent crude protein. Milk production of dairy cattle was not altered when a diet consisting of 22.5 percent bagasse, 20 percent molasses, and 57.5 percent concentrate was substituted for a conventional concentrate-sorghum silage diet (Randel, 1970b). Replacement of cottonseed hulls with bagasse in a 25 percent roughage diet for lactating dairy cattle had no detrimental effect on milk production (Marshall and Van Horn, 1975).

Clanton and Harris (1966) found that calves fed ensiled sugar beet tops that had been field-wilted achieved higher daily gains than calves fed unwilted ensiled beet tops or calves that were pastured on beet tops (0.61, 0.49, and 0.54 kg/day, respectively). In another study (Rush, 1977), average daily gain decreased linearly as beet tops were substituted for corn silage in a growing calf ration. Calves went from an average gain of 0.88 kg/day with no beet tops to 0.59 kg/day when beet tops were substituted for 60 percent of the corn silage. Levels of beet tops above 30 percent of the diet produced scouring. When pasturing cattle on beet tops, one hectare can carry one animal unit for 36 days with animal gain approximately 0.5 kg/day (Leonard et al., 1973). Sugar beet tops are best utilized when combined with another residue (such as cornstalks) or forage.

Sugarcane tops are a fibrous feed that is normally discarded in the field during the harvesting of sugarcane. Growing cattle fed fresh tops for 112 days maintained their weight, while those fed ensiled tops lost 0.27 kg/day (Estima et al., 1967). The addition of molasses or cassava roots had little effect on gain or feed consumption, but supplementing 1.5 kg/day of cottonseed meal stimulated both intake and daily gain. In another study (Pate et al., 1971), growing steers full-fed on tops were unable to maintain their weight, losing 0.1 kg/day. Subsequent supplementation with 0.9 kg/day of cottonseed meal increased gains to 1.1 kg/day and increased dry-matter intakes of cane top from 5.9 to 7.4 kg/day. Finishing cattle were fed cane tops at 0, 17, 34, and 51 percent of the diet (Pate and Coleman, 1975). Steers fed the 17 percent level gained as well as those receiving the control diet (0.9 kg/day), and feed efficiency was similar.

Residues from sunflower seeds and faba bean production appeared to be quite high in digestibility (70.5 and 61.4 percent, respectively). Their feeding value would be expected to be quite high on the basis of these laboratory digestibilities, but animal growth data are unavailable (Kernan et al., 1980).

Vegetable and fruit field residues consist of leaves, vines, culls, and unharvested crops. Thus, unlike other crop residues, which are almost universally low-quality roughages, these residues vary widely in their nutritional value and in their availability. The University of California summarized its previously unpublished research in the area of unusual feedstuffs for livestock (Leonard et al., 1973), from which much of the following information was taken.

Lettuce contains approximately 93 percent moisture. On a dry basis it is comparable to some cereal grains, with a TDN of 70 percent and 7 percent digestible protein. Cattle prefer to consume the heads rather than the leaves. Cattle consumed up to 16 kg carrots/day in addition to supplement and gained about 0.1 kg/day. Fresh peaches can be fed at a rate of 9–14 kg/day, with no evidence of scouring. When 2.7 kg dried peaches were fed, scouring ensued, and shortly thereafter the animals refused the fruit. One-half this level was consumed for 10 days with no ill effects. Pears were consumed at a lower rate, and spoiled pears were more readily refused than spoiled peaches. Scouring also occurred with fresh prunes at levels over 6.8 kg/day. Dried prunes were fed at the level of 2.7 kg/day with no detrimental effects. Grapes have been fed to cattle at levels up to 16 kg/day and raisins at 2.7 kg/day with no ill effect. Consumption of peaches created a slight off flavor in milk.

By-products from sweet corn production are available and of quite high value. The dry-matter digestibility of the stalks remaining after harvest of the ears is over 65 percent. Up to 20 percent of the ears are not harvested because of weather and other factors. The whole-plant material is over 66 percent digestible. Cannery residue is over 70 percent digestible. All of the products are high in moisture but also high in nutritive value, nearly equal to field corn silage.

Pea vines are also high in nutritional value (63.4 percent TDN). They are collected at central locations and present a disposal problem.

Leaf waste from cauliflower, a product of the packing shed, was separated to provide a poultry meal fraction containing 375 to 620 mg/kg xanthophyll and 26 to 31 percent protein, and a cattle meal fraction that contained 17 to 21 percent protein (Livingston et al., 1972). The xanthophylls were equivalent to dehydrated alfalfa meal as a pigmenting agent for broiler skin. Sweet potato vine meal has also been shown to be effective as a pigmenting agent (Garlich et al., 1974).

Haines et al. (1959) fed celery tops to Hereford steers at 0, 10, 20, and 30 percent of a diet containing sorghum, soybean meal, citrus pulp, and citrus molasses for 112 days. Diets were isonitrogenous, and dehydrated Bermudagrass was fed ad libitum. Average daily gains were 1.01, 1.03, 0.85, and 1.12 kg/day, respectively. Steers receiving the 20 percent celery top ration were the least efficient converters of concentrate and roughage. A digestion trial with four steers yielded a TDN of 79.5 percent. Crude protein was 25.0 percent on a dry-matter basis.

Cassava, although not grown widely in the United States, is a major food crop in many areas of the tropics and could assume importance as both a human and animal food, since its yield of energy per unit land can be quite high. Because of the presence of a toxic glucoside, cassava leaves must be dried or boiled prior to feeding to reduce toxin levels. Cassava leaves have a high protein content (25 percent), although they may be deficient in methionine and tryptophan (Rogers and Milner, 1963). Use of cassava leaves in poultry diets results in poor growth unless supplemented with methionine (Ross and Enriquez, 1969).

PROCESSING METHODS

If crop residues are to be used to meet the energy requirements of growing and lactating ruminants, their feeding value must be increased. There seem to be two possibilities for this at the present time. One is manipulation of harvest time or genetics to obtain higher-quality residues, and the other is to treat residues physically or chemically to increase digestibility and/ or intake.

Corn

Many chemicals have been screened in laboratory experiments for their potential to enhance digestibility. However, only four are being routinely used in experimentation with animals: sodium hydroxide, ammonia, calcium hydroxide, and potassium hydroxide.

Chemical treatment breaks the ether linkages between lignin and cellulose or hemicellulose. The saponified lignin is not soluble in acid and therefore may be measured by most methods (Lau and Van Soest, 1981). However, the resultant increase in digestibility is still obtained.

Modes of action for chemical treatment of crop residues have been described by Waller (1976). Chemical treatment solubilized some of the hemicellulose while not changing the cellulose content. Extent of bacterial digestion in vitro was increased for both cellulose and hemicellulose. Chemical treatment, especially with sodium hydroxide, increased the rate of digestion for both cellulose and hemicellulose. Therefore, one could conclude that the modes of action of chemical treatment, especially treatment with sodium hydroxide, include (1) solubilization of hemicellulose, (2) increasing the extent of cellulose and hemicellulose digestion, and (3) increasing the rate of cellulose and hemicellulose digestion, possibly by swelling.

During the past several years two systems for application of chemical treatments primarily with sodium hydroxide, have evolved (Jackson, 1977). One system, described by Rexen and Thomsen (1976), involves application of a concentrated sodium hydroxide solution prior to pelleting. Heat produced in pelleting causes the chemical reaction to go rapidly to completion. The process involves collection and transportation of residue to a central processing plant, followed by chopping or grinding and mixing concentrated sodium hydroxide solution with the residue. Sodium hydroxide serves as a pellet binder in addition to its effect on nutrient digestibility. A dense, hard pellet is then produced; excess moisture is removed in the cooling process.

Response to pelleting of cornstalks is shown in Table 50. As indicated by the in vitro digestibility values, the stalks were of very low value. Both NaOH treatment and pelleting increased rate and efficiency of gains. The combination of pelleting and NaOH treatment was slightly more than additive. The diets contained one-half relatively low-quality alfalfa as well. While animal performance was not especially good, feeding the treated

TABLE 50 Pelleted and Sodium Hydroxide-Treated Cornstalks[a]

Performance	Chopped Control	Pelleted Control	Chopped NaOH Treated	Pelleted NaOH Treated
Daily gain (kg)	0.16	0.28	0.24	0.31
Daily feed (kg)	6.3	6.5	6.6	5.4
Feed/gain ratio	0.026	0.043	0.036	0.057
IVDMD[b]	37.0	38.0	59.0	61.0

[a]Twenty-two calves/treatment, 205 kg initial weight, 91-day trial. Diets were 50 percent alfalfa (50 percent IVDMD).
[b]In vitro dry-matter disappearance.

SOURCE: Klopfenstein, unpublished data.

pellet diet approached economic feasibility. Pelleting (with sodium hydroxide) better-quality stalks and feeding with higher-quality alfalfa may increase the economic feasibility. The primary advantage of pelleting is that the product can be stored and transported easily and will be consumed in large quantities by livestock because of its high bulk density. The major disadvantage is the cost of collection and transportation of bulky residues prior to treatment and pelleting.

The second process is an on-the-farm method that involves collecting residues in the field behind the combine with a bunch wagon or with a stack wagon or baler. Following grinding or harvesting with a forage harvester, the residue is transported to the treatment area, mixed in a mixer wagon with the chemical, and sufficient water added to raise the moisture to 15 to 65 percent. Small-scale, one-step machines that handle small, square straw bales are marketed in Europe for this purpose. The material is then ensiled or fed after 24 to 48 hours of reaction. Results of studies with this type of treatment are encouraging. Steers gained over twice as rapidly and required approximately half as much feed per unit of gain with 4 g/100 g dry-matter sodium hydroxide treatment of corncobs (see Table 51).

Table 52 summarizes three trials comparing performance of calves fed corn silage and chemically treated husklage. The husklage was treated with 3 g sodium hydroxide and 1 g calcium hydroxide/100 g dry matter and was fed as 80 percent of the diet with 20 percent supplement. Corn silage was fed as 90 percent of the diet with 10 percent supplement. The calves fed both corn silage and treated husklage gained 0.75 kg/day. Daily feed varied only slightly; therefore, feed/unit of gain was similar.

Cornstalks were harvested with a forage harvester equipped with a stalker head at two stages of maturity (Berger et al., 1979). At each stage of harvest one-half of the stalks were chemically treated. The average response to chemical treatment was less than 10 percent improved feed efficiency. Unknown factors may be influencing the ability to accomplish large-scale treatment.

Several researchers have shown that animal response to sodium hydroxide treatment (digestibility or efficiency of gains) is less than laboratory predictions (Klopfenstein et al., 1972; Ololade et al., 1970; Rexen and Thomsen, 1976; Singh and Jackson, 1971).

Ruminally fistulated lambs were used to measure the effect of increasing levels of sodium hydroxide treatment on rate of fiber passage and rate of ruminal fiber digestion (Berger et al., 1980). In vitro dry-matter digestibility increased from 45.1 percent for the control to 83.1 percent for the 8.0 g/100 g dry-matter sodium hydroxide diet. As the level of sodium hydroxide treatment increased, the in vivo rate of passage increased lin-

TABLE 51 Performance of Steers Fed Sodium Hydroxide-Treated Corncobs

Performance	Control	4 Percent Sodium Hydroxide[a]
Number of steers	5	15
Initial weight (kg)	205.0	203.0
Daily gain (kg)	0.30	0.73
Daily feed (kg)[b]	4.1	5.4
Feed/gain ratio	14.3	7.5

[a]Corncobs constituted 80 percent of the diet.
[b]Dry basis.

SOURCE: Koers et al. (1970).

early. Rate of ruminal fiber digestion was measured using nylon bags containing 0.15 g cotton fiber. As level of sodium hydroxide in the diet increased, the rate of ruminal cotton digestion decreased linearly (r^2 = .93). These data suggest that sodium intake, the effect of sodium hydroxide on plant-fiber digestion, and possibly a chloride anion deficiency affect rate of fiber passage and digestibility.

Another possible effect of high sodium intakes in ruminants is on mineral metabolism. If sodium load causes a mineral imbalance, then supplementation of other minerals may be necessary. Lambs were fed ad libitum either untreated corncobs, 4 g/100 g dry-matter sodium hydroxide-treated corncobs, 4 g/100 g dry-matter sodium hydroxide-treated corncobs

TABLE 52 Treated Husklage Versus Corn Silage[a]

Performance	Corn Silage	Treated Husklage[b]
Daily gain (kg)	0.75	0.75
Daily feed (kg)[c]	7.26	7.17
Feed/gain ratio	9.62	9.60
Feed cost (¢/kg)	9.35	7.7
Cost of gain (¢/kg)	90.0	74.0

[a]Summary of 3 trials, 96 calves/treatment.
[b]Treated with 3 g sodium hydroxide and 1 g calcium hydroxide/100 g dry matter.
[c]Dry-matter basis.

SOURCE: Klopfenstein (1978).

plus supplemental minerals, or 3 g/100 g dry-matter sodium hydroxide: 1 g/100 g dry-matter potassium hydroxide-treated corncobs plus supplemental minerals (Paterson et al., 1978b). Supplemental minerals were supplied in the diet by weight to provide ratios of 1 Na:1 K, 2 Na:1 Ca, 2 Na:1 Cl, and 6 Na:1 Mg. No difference in sodium balance was apparent among treatments at the end of 21 days, but all animals fed the sodium hydroxide-treated cobs were in a negative sodium balance the first 14 days before returning to a positive balance after 21 days. Calcium, magnesium and phosphorus balance were not adversely affected either by sodium load or by supplementation with excess minerals. However, it did appear that supplemental potassium and chloride may be necessary to prevent negative balances.

Information on the application of ammonia (either as anhydrous gas or aqueous liquid) to crop residues indicates that it may also improve residue feeding quality. Ammonia appears to react in a manner similar to sodium hydroxide. However, the reaction time is much longer (up to 20 days) (Waiss et al., 1972) than with sodium hydroxide treatment (24 hours), and the residue must be stored in an airtight structure so there will be no loss of ammonia. The two major advantages of using ammonia are (1) the use of the residual nitrogen as a nonprotein nitrogen source in the diet and (2) no mineral residue remaining that might be detrimental to the animal or to the soil to which the manure is added. Research has shown that animals will not eat ammonia-treated residues unless the residues are aerated or mixed with a fermented feed so that the organic acids neutralize the ammonia (Waller, 1976).

At ad libitum intakes the dry-matter digestibility of cornstalks was increased from 36.8 percent for sheep fed untreated stalks to 47.0 percent for sheep fed stalks with 2 g ammonia/100 g dry matter. The addition of 3 or 4 g ammonia/100 g dry matter did not further improve digestibility. The intake of the aerated stalks increased from 398 g to 997 g/day for lambs fed untreated stalks and 4 g ammonia/100 g dry-matter stalks, respectively (Paterson, 1979). Cattle performance on ammonium-treated cobs by the on-the-farm method is shown in Table 53. In this case ammonium hydroxide-treated cobs were fed in combination with calcium hydroxide-treated cobs. Ammonium treatment appeared to produce increased performance over calcium hydroxide treatment alone. Efficiency of gains produced by the mixture was superior to that of the 4 g sodium hydroxide-treated cobs. This would indicate that the ammonium hydroxide-treated cobs were quite efficiently utilized. Intake was somewhat lower with ammonium hydroxide-treated cobs, probably because of free ammonia that still remained in the cobs at feeding time. While ammonia treatment requires a closed system, there are distinct advantages to its use.

TABLE 53 Performance of Growing Calves Fed Different Chemically Treated Cobs

NaOH:CaOH[a]	Average Daily Gain (kg)	Average Daily Feed (kg)[b]	Feed/Gain Ratio
4:0	1.17[c]	10.10	8.6
3:1	1.34[d]	10.32	7.7
2:2	1.14[c]	9.87	8.7
1:3	1.27[d]	9.97	7.8
0:4	0.97[e]	9.04	9.4
1/4 NH$_4$[f]	1.16[d]	8.82	7.6

[a] Sodium hydroxide:calcium hydroxide = 1 g each/100 g cob dry matter.
[b] Dry matter.
[c,d,e] Means in the same column with different superscripts differ significantly ($P < .05$).
[f] 4 g ammonium hydroxide and 4 g calcium hydroxide-treated cobs mixed 1:1 at feeding time.

SOURCE: Waller and Klopfenstein (1975).

Calcium hydroxide is less caustic than sodium hydroxide, and greater amounts may be necessary to obtain performance equal to sodium hydroxide treatment alone. Since calcium is not metabolized in the same manner as sodium, the increased levels may not be as detrimental as a high sodium load.

Digestion studies with corncobs, cornstalks, and wheat straw indicated that moisture level affected response to calcium hydroxide treatment (Paterson et al., 1980). Forty percent moisture was better than 20 or 60 percent moisture. Sixty percent residues fermented, while treated residues with 20 percent moisture appeared to have much of the calcium hydroxide unreacted. Even though a 40 percent moisture level increased digestibilities and intake, the treated residues showed signs of mold growth after 7 days.

In an in vitro study, cobs treated with 5 g calcium hydroxide/100 g dry matter did not reach maximum digestibility for 10 days but cobs treated with 1 or 2 g sodium hydroxide replacing calcium hydroxide reached maximum digestibility by days 2 and 4, respectively (see Figure 12). It appears that in a practical system the combination of 1 g sodium hydroxide and 3 or 4 g calcium hydroxide/100 g dry matter could be used at 60 percent moisture, and the chemical dilignification reaction would proceed before the bases were neutralized through microbial fermentation. A cattle growth trial using corncobs demonstrates this by showing much greater response to cobs treated with 1 g sodium hydroxide plus 3 g calcium hydroxide than to cobs treated with calcium hydroxide alone (Table 53).

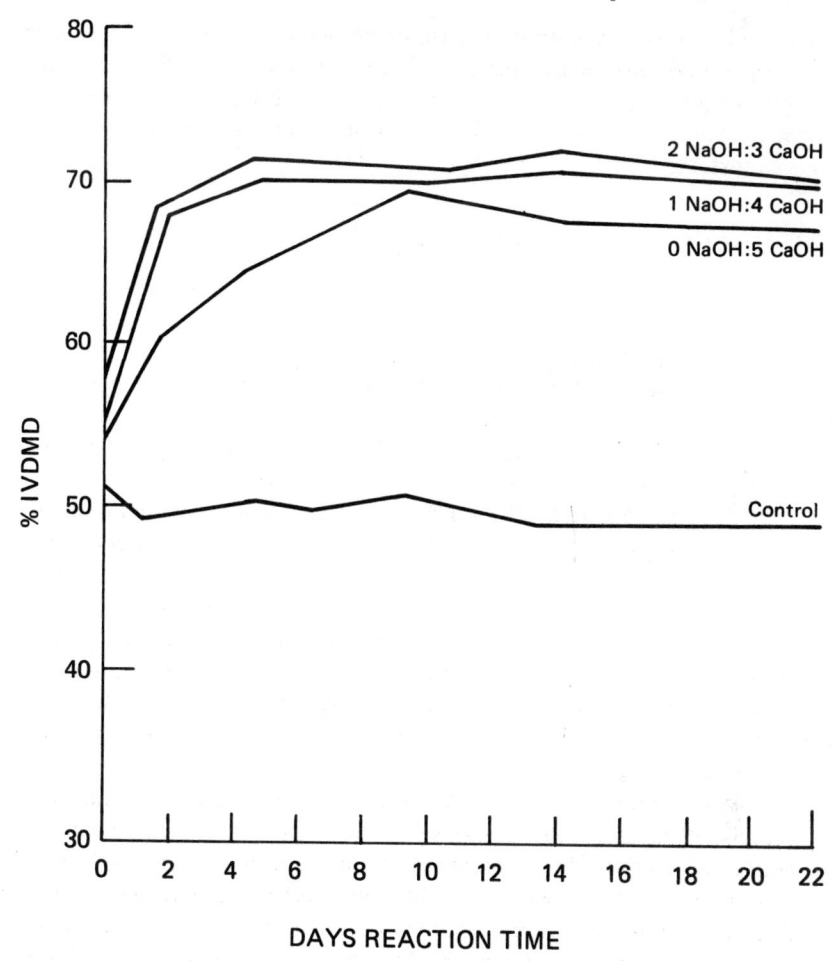

FIGURE 12 In vitro dry-matter-disappearance (IVDMD) for cobs with different combinations of sodium hydroxide (NaOH) and calcium hydroxide (CaOH). SOURCE: Paterson et al. (1980).

Another system for increasing digestibility of crop residues is the use of high pressure steam with or without added chemicals (Bender et al., 1970; Guggolz et al., 1971; Klopfenstein et al., 1974; Klopfenstein and Bolsen, 1971; Umunna et al., 1972). In this system the fibrous material is moistened and placed in a pressure vessel; steam pressures of 14 to 28 kg/cm^2 are applied for a few seconds and then the pressure is released. The reaction is a mild acid hydrolysis because some organic acids are produced in the reaction. Hemicellulose is solubilized, and probably the digestibility of the cellulose is also increased. This process has been developed commercially using a continuous-flow digester (Bender et al., 1970). Digestibility of corncobs was increased by 16 percentage units through steam hydrolysis, and daily gains of growing calves were doubled (Klopfenstein, 1975).

Wheat and Other Small Grains

Straws from small-grain production tend to be bulkier and are lower in quality than corn residues. Response to pelleting and sodium hydroxide treatment have been quite good (Coombe et al., 1979a; Garrett et al., 1979; Rexen and Thomsen, 1976). Numerous commercial plants in several European countries are producing treated straw pellets.

The mode of action and compositional changes caused by sodium hydroxide treatment are similar to those for corn plant residues (Coombe et al., 1979b; Lesoing et al., 1980b). Pelleting and sodium hydroxide treatment each increased rate and efficiency of gains and the effects were essentially additive for both wheat and barley straws (Coombe et al., 1979a,b). Gains of over 0.6 kg/day were obtained with treated, pelleted straws plus protein supplement. Garrett et al. (1979) obtained similar results with rice straw.

Straws can also be treated by an on-the-farm system similar to that used for corn residues. Wheat straw treated by this method showed increased digestibility (Hasimoglu et al., 1969; Jared and Donefer, 1970; Lesoing et al., 1980b) and produced increased rate and efficiency of gains (see Table 54). Because of the low protein content of crop residues, the cost of supplemental protein is very important. Nonprotein nitrogen as the only supplemental nitrogen has limited value in treated-residue rations. These same principles were demonstrated with oat straw (Donefer et al., 1969; Saxena et al., 1971). Barley straw (Jayasuriya and Owen, 1975; Mowat and Ololade, 1970; Wilkinson and Santillana, 1978) as well as ryegrass straw (Anderson and Ralston, 1973) shows increased digestibility with chemical treatment.

Wheat straw was treated, fed to growing calves, and compared with corn silage and untreated straw (see Table 55). Treatment increased the rate and efficiency of gain compared with untreated straw in diets containing 50 percent straw (Lesoing et al., 1980a). Adding potassium, calcium, chlorine, and magnesium to diets containing 80 percent treated straw also increased rate and efficiency of gains. This demonstrates that the sodium residue creates a nutritional problem that can be at least partially solved. Calves fed treated straw showed reasonably good performance, but performed considerably less well than they had on diets of corn silage.

TABLE 54 Performance of Lambs Fed Wheat Straw

Ration[a]	Daily Gain (g)	Daily Feed (g)[b]	Feed/Gain Ratio	Organic Matter Digestibility (%)[c]
Straw + SBM[d]	36	908	25.0	49
Straw + urea	−45	636	—	51
4% NaOH straw + SBM	160	1226	7.6	60
4% NaOH straw + urea	82	999	12.4	59

[a]Diets were 70 percent straw and 30 percent supplement (soybean meal, urea, wheat, grain, minerals, and vitamins).
[b]Dry basis.
[c]Digestibility of straws assuming 90 percent digestibility of the supplements.
[d]Soybean meal.

SOURCE: Hasimoglu et al. (1969).

Treatment of straws for use in beef cow gestating diets shows some promise. Acock et al. (1979) showed that cows gained weight on treated wheat straw plus protein. Ammonia treatment of straws could be accomplished without grinding and mixing (Garrett et al., 1979; Sundstol et al., 1978).

TABLE 55 Effect of Treatment of Wheat Straw and Balancing Minerals for High Sodium Intake on Rate and Efficiency of Gain of Steers

Treatment[a]	Daily Gain (kg)[b]	Daily Feed (kg)[c]	Feed/Gain Ratio
Corn silage	1.05	6.14	5.79
50% Untreated wheat straw	0.62	5.53	9.05
50% Treated wheat straw + minerals	0.74	5.83	8.04
80% Treated wheat straw, no minerals	0.54	5.32	9.94
78% Treated wheat straw + minerals	0.65	5.51	8.44

[a]Wheat straw was treated with 3.15 g sodium hydroxide and 1.19 g potassium hydroxide/100 g wheat straw dry matter.
[b]Steers were weighed after an overnight shrink on day 109. Steers were fed an equal amount of a standard corn silage diet on days 103 through 108.
[c]Dry-matter basis.

SOURCE: Lesoing et al. (1980a).

Gestating beef cows were fed 2.7 kg alfalfa hay/day and wheat straw ad libitum (Faulkner et al., 1981). Ammonia-treated bales were consumed in greater quantities than control, liquid supplement, or NaOH-treated bales. Weight gains were 0.4 kg/day on the ammonia-treated straw.

Waagepetersen and Thomsen (1977) have shown that the rate of reaction of ammonia is increased with increasing temperature. They have developed equipment to be used for on-farm treatment at elevated temperatures.

A lamb digestion trial and two lamb growth trials were conducted to compare straw treated with ammonium hydroxide (6 g/100 g straw dry matter), calcium hydroxide (4 or 5 g/100 g straw dry matter), an ensiled mix of 2 g calcium hydroxide and 3 g ammonium hydroxide, or a 50:50 combination of 4 g calcium hydroxide:6 g ammonium hydroxide mixed at feeding time (Asadpour, 1978). Moisture content of straw was 65 percent. Table 56 lists dry-matter, neutral (NDF), and acid-detergent fiber (ADF) digestibilities for the trial. Hydroxide treatments improved all digestibilities over the untreated control. However, lambs fed the straw treated with ammonia or the hydroxide combinations averaged greater digestibilities than lambs fed straw containing either 4 or 5 g calcium hydroxide alone. There were no differences between lambs fed straw treated with 6 g ammonium hydroxide or either of the calcium-ammonium combination mixes.

Performance data from two growth trials were pooled and are presented in Table 56. Lambs fed straw diets containing the hydroxides had better daily gains and feed conversions than those fed the untreated control. Lambs fed straw containing only ammonia or a calcium-ammonium mix gained better than did lambs fed straw treated with calcium hydroxide alone. While there did not appear to be any difference in digestibility between the 4 g and 5 g calcium hydroxide-treated diets, there did appear to be an advantage in average daily gain with lambs fed 5 g calcium hydroxide.

Pressure treatment with steam has also increased wheat straw digestibility (Umunna and Klopfenstein, 1972).

Soybean

Soybean residues are very high in lignin. Being legumes, they do not respond well to chemical treatment. Little processing research has been conducted (Bottje et al., 1979).

Grain Sorghum

Grain sorghum residue responds to sodium hydroxide and calcium hydroxide treatment (Chandra and Jackson, 1971; Koers et al., 1972; Sherrod

TABLE 56 Effect of Chemical Treatment of Wheat Straw on Lamb Dry Matter and Neutral Detergent and Acid Detergent Digestibility and Growth

Treatment[a]	Dry Matter Digestibility (%)	Neutral Detergent Digestibility (%)	Acid Detergent Digestibility (%)	Average Daily Gain (g)	Daily Dry-Matter Intake (g)	Gain/Feed Ratio
Control (plain straw)	47.6	50.5	45.6	8.6	663	0.014
6 g ammonium hydroxide/100 g straw	55.8	63.1	57.6	98	863	0.113
2 g calcium hydroxide + 5 g ammonium hydroxide/100 g straw	56.5	61.6	57.1	75	821	0.091
2 g calcium hydroxide + 5 g ammonium hydroxide/100 g straw mixed at feeding time	54.6	59.4	55.1	97	867	0.112
5 g calcium hydroxide/100 g straw	51.1	56.9	50.4	80	858	0.093
4 g calcium hydroxide/100 g straw	53.6	55.2	51.7	60	808	0.074

[a]Straw diets were supplemented with protein, Ca, P, salt, and vitamins and fed as 80 percent straw + 20 percent supplement at 90 percent of ad libitum intake. Straw was mixed with chemicals and water (40 percent dry matter) and ensiled 21 days before feeding.

SOURCE: Asadpour (1978).

and Summers, 1974) but probably not as well as corn residue or straws. Pelleting did not improve performance (Bolsen et al., 1977) but pelleting along with sodium hydroxide treatment has not been attempted. Little or no research has been conducted with ammonia treatment.

Other Residues

A feedlot trial with raw and ammoniated rice hulls (approximately 1.5 g/100 g dry matter) replacing sorghum at 3, 6, and 9 percent of the diet resulted in no effect on gains, feed efficiency, or carcass characteristics (Tillman et al., 1969). Ammoniation had no effect on organic-matter digestibility or nitrogen retention. Hutanuwatr et al. (1974) increased the in vitro dry-matter digestibility of rice hulls from 22.5 to 40 percent by adding 12 g sodium hydroxide/100 g dry matter followed by washing. Washing was superior to acid neutralization but was accompanied by a 16.6 percent dry-matter loss. McManus and Choung (1976) reported that high levels of sodium hydroxide resulted in extensive solubilization and removal of silica and lignin. Acid-neutralized sodium hydroxide-treated hulls (0, 2.5, 5, 10, and 15 g sodium hydroxide/100 g dry matter) were fed to sheep for 91 days with an equal amount of alfalfa pellets. Sheep receiving rice hulls treated with the three highest levels of sodium hydroxide maintained or gained weight, while sheep on the lower two levels lost weight (Choung and McManus, 1976). Only the 15 g sodium hydroxide-treated hull diet had significantly greater organic-matter digestibility than the control in an accompanying metabolism trial (52 versus 42 percent). Acid-detergent-fiber digestibility was greatly increased (62 versus 13 percent). Sheep tolerated the high sodium levels, although water consumption increased. An alternative treatment was employed by Daniels and Hashim (1977). Fungal cellulases were added at various levels to a maximum of 1,250 mg/kg. In vitro dry-matter digestibility was highest with 375 mg/kg (30.6 versus 17.3 percent with no added cellulase). In a digestion trial with Holstein steers, addition of cellulase at levels of 250 to 2,000 mg/kg to rice hulls yielded increases in dry-matter, energy, and protein digestibility.

Barton et al. (1974) treated peanut hulls with seven compounds: ammonia, sodium hydroxide, a combination of ammonia and sodium hydroxide, calcium hypochlorite, chlorine, dioxane, and dimethylsulfoxide. Only calcium hypochlorite appreciably improved in vitro dry-matter digestibility (40 versus 25 percent).

Chemical treatment of sugarcane bagasse has dramatically increased its value as a feed for ruminants. Randel et al. (1971) fed lactating dairy cattle 40 percent bagasse diets (raw, and treated for 24 hours with sodium hydroxide), and a control standard diet. Milk production was greater for the control and treated bagasses, than for the raw bagasse (16.5, 17.2, and 12.5 kg/day, respectively). In a digestion trial conducted at restricted intake with bagasse at 40 percent of the diet, TDN was increased from 55.8 to 67.4 when the bagasse was treated with 2 g sodium hydroxide/ 100 g dry matter (Randel, 1972). Bagasse silage, composed of bagasse, molasses, urea, and whole corn, was more acceptable to crossbred steers when the bagasse was treated with 5 g sodium hydroxide/100 g dry matter (Andreis and DeStefano, 1978). In a 182-day trial, average daily gain was 0.71 kg with the treated bagasse as opposed to 0.44 with raw bagasse. Feed conversion with treated bagasse was 12.3 versus 18.7 for the untreated bagasse. Martin et al. (1974) employed both calcium hydroxide and sodium hydroxide at rates ranging from 0 to 16 g/100 g bagasse. Both in situ and in vitro digestion increased with increasing levels of treatment, but sodium hydroxide was superior to calcium hydroxide. Steam-pressure treatment also increased in vitro digestibility from 28.3 at 4 atm. for 15 minutes to 44.9 at 6 atm. for 30 minutes. In a subsequent study the combination of sodium hydroxide treatment with steam-pressure treatment resulted in the largest increase in digestibility (Martin et al., 1976).

UTILIZATION SYSTEMS

Use of low-quality crop residues has been restricted primarily to ruminants on maintenance rations, such as gestating beef cows. In many parts of the United States, cornstalks and milo stubble can be grazed. This is the most economical system for the use of crop residues (Ward, 1978). There is essentially no energy (fuel) cost and no machinery cost, and nothing of value is removed from the soil. The manure deposited by the cows adds indigestible fiber to the soil, and if the cow is supplemented, more nitrogen and minerals are deposited than were in the crop residue consumed. Grazing of cows on crop residues does require good management to obtain the most efficient production.

In the Midwest, spring-calving cows are in midgestation during peak stalk or stubble grazing. Calves are generally weaned prior to this time so that the cows' requirements are only slightly above maintenance. Thus, the cows are grazing stalks at the time of lowest nutrient requirements and grazing grass during peak requirements of lactation and breeding. Two problems exist with this system, however. First, the total feed supply is available on the first day cows are allowed to graze stalks or stubble, but because of weathering and other losses and selective grazing, the nutritional value of feed available declines with time. Concurrently, the nutrient requirements are increasing because of fetal growth. The second problem is the period of time from the end of residue grazing until green grass is available. This hiatus generally occurs during the calving period when the nutrient requirements of the cow are rather high. A harvested feed is necessary. There is an opportunity for increased use of crop residues during this time period, but the nutritional value must be higher than that of present material to meet the needs of the cow.

Another important problem with the grazing of stalks or stubble is the weather. In higher-rainfall areas of the eastern United States the residues deteriorate rapidly, and muddy conditions may prevent continual grazing. In the western Corn Belt and Plains states, adequate grazing can usually be obtained. Snow cover may prevent grazing, however, and some harvested forage is needed. Harvested crop residues can meet this need rather economically. Ward (1978) reported an average of 88 days of crop residue grazing in eastern Nebraska. An average of 250 kg/cow of supplemental stacked residue was fed as well. Cows gained over 0.25 kg/day and were supplied 1 ha corn or milo residue/cow.

Feeding straws to cows is of interest because of the availability of wheat straw in many areas. Other residues (corn and grain sorghum) may not be available in these areas.

Beef cow/calf production systems are illustrated in Table 57. These systems range from extensive, using grass, to intensive, using only crop residues. Any of these systems can be economical depending upon management and cost of grain and protein.

A system of growing and finishing beef calves is shown in Table 58. The calves were allowed to graze cornstalks followed by harvested stalklage feeding. Summer grazing of grass preceded a high-grain finishing period. Less than 800 kg grain were needed to finish these animals, about

TABLE 57 Cow-Calf Production

Condition	Extensive	Intensive Corn Residue[a]	Intensive Straw[a]
Mid gestation	Stalk grazing	Stalklage[b]	Treated straw
Late gestation	Stalklage[b,c]	Stalklage[b,c]	Treated straw[c]
Early lactation	Grass	Treated stalklage[d]	Treated straw[d] and concentrate
Mid and late lactation	Grass	Treated stalklage[d]	Treated straw[d]

[a]Requires 4 metric tons residue/year and 170 kg protein.
[b]Could be replaced by treated straw.
[c]Requires .04 kg protein/kg residue.
[d]Requires .06 kg protein/kg residue.

TABLE 58 Beef Production Systems With Heifers Utilizing Crop Residues

Cornstalk grazing, 56 days[a]	0.52 kg/day
Harvested stalklage, 75 days[b]	0.02 kg/day
Summer grass, 110 days[c]	0.70 kg/day
High-grain finishing, 93 days[d]	1.34 kg/day

[a]Equivalent of 280 kg residue, 0.06 kg protein needed/kg crop residue.
[b]450 kg residue.
[c]Equivalent of 770 kg residue.
[d]Equivalent of 1,600 kg residue.

SOURCE: Faulkner and Ward (1981).

1.7 kg/kg beef produced. Forages were emphasized, especially crop residues, and compensatory gain was exploited.

Alfalfa, which is a good source of supplemental protein and minerals, may be the logical choice to add to sodium hydroxide-treated rations in an attempt to slow the rate of fiber passage, increase the extent of rumen fiber digestion, and equilibrate mineral balance (Paterson et al., 1978a). The additions of various levels of chopped alfalfa hay in sodium hydroxide-treated corncob or cornstalk diets were evaluated in lamb digestion trials.

Dry-matter digestibility of the diets was improved from 50 percent for the all-alfalfa diet to 63 percent for the 50:50 alfalfa:NaOH-treated cob diet. However, when the sodium hydroxide-treated cobs composed more than 50 percent of the diet, dry-matter digestibility decreased to approximately 60 percent on the all-treated-cob diet. Lambs exhibited large individual fluctuations in digestibility with diets of 100 percent sodium hydroxide-treated cobs. This variation and the decreased digestibility may be a function of sodium load and rate of fiber passage. In vitro dry-matter-disappearance values compared well with the in vivo digestibilities in diets of up to 50 percent treated cobs. The difference in digestibilities for the two diets was approximately 20 percentage units at the all-treated-residue diet. The failure of the in vivo values to correspond with the expected potential digestibility of in vitro values seems to suggest the effects of sodium on rate of passage and fiber digestion. Similar results were found with cornstalks.

Hydroxide-treated corncobs were fed along with alfalfa hay to growing steers to evaluate (1) response to chemical treatment and (2) associative effects of alfalfa and hydroxide-treated cobs (Paterson et al., 1978a). Treatment of corncobs with 3 g sodium hydroxide and 1 g calcium hydroxide/100 g dry matter increased gains of growing calves an equivalent of 0.27 kg/day. Feeding 50 percent alfalfa with treated cobs increased the response to chemical treatment and showed positive associative effects (see Figure 13).

Jackson (1978), in an excellent review, has summarized the effects of diluting sodium hydroxide-treated straws with either concentrates or high-quality forages. Concentrates dramatically reduced the digestibility of the fiber, especially at higher levels. Below 15 percent of the diet, concentrates seemed to have little effect on fiber digestion. High-quality forages, as was shown here with alfalfa, did not reduce digestibility of crop-residue fiber. Obviously, the formulation of the complete diet is very important in realizing the full benefits of crop-residue treatment.

POTENTIAL UTILIZATION

Crop residues will likely play a very large role in future production of meat and milk by ruminants. Low-cost extensive systems presently make good use of some residues but meet the needs of only a small portion of

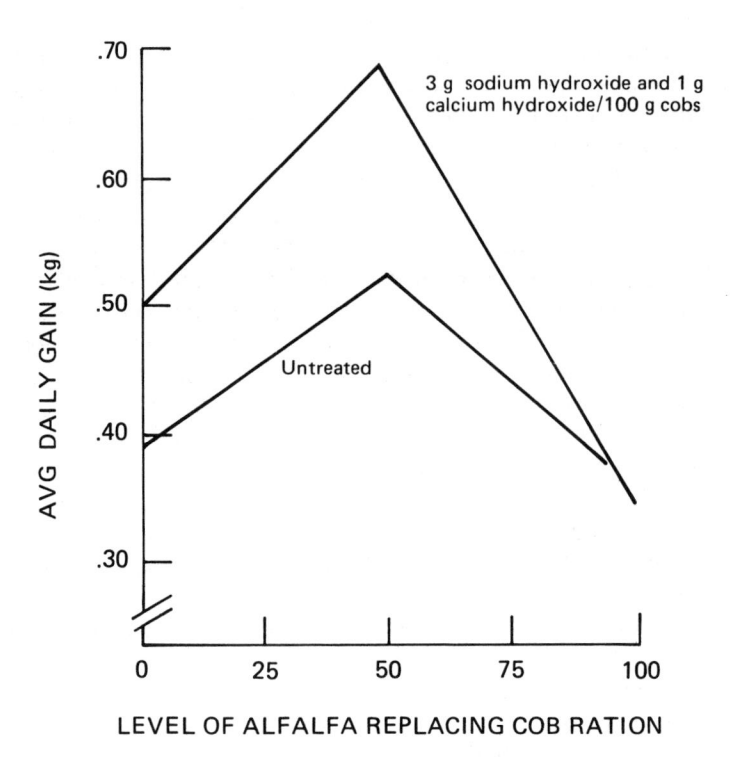

FIGURE 13 Average daily gains of steers fed corncob ration with 0, 50, or 100 percent alfalfa hay addition. SOURCE: Paterson (1979).

the ruminant population. Utilization of crop residues can be increased by cultural and management practices, such as stage of maturity at harvest.

The greatest increase in utilization of crop residues will likely come from chemical treatment in combination with improved management practices. Sodium hydroxide has been widely studied as a chemical for crop-residue treatment. While the treatment is effective, concerns about human safety and sodium residues may limit its ultimate usefulness. Probably treatments with ammonia and combinations of chemicals will prove more useful.

In addition to feeding systems, systems of harvest, treatment, and storage can be developed to optimize the use of residues.

ANIMAL AND HUMAN HEALTH PROBLEMS AND REGULATORY ASPECTS

There are no special animal or human health problems involved with feeding crop residues or treated residues to ruminants. Any residue stored improperly will likely mold and can potentially produce aflatoxins that can have devastating effects on animals.

Most crops that provide residues for feeding have had herbicides or insecticides applied to them. These compounds present no health problems to animals or people as long as label directions are adhered to strictly. Compounds that are cleared for use on crops for silage will obviously be suitable for use when crop residues are harvested. Herbicides are generally not applied near harvest, but on some occasions insecticides might be applied shortly before harvest. The insecticide label will indicate safety for use on forage crops.

For some crops, such as cotton, the whole plant is never intended for silage, and clearance for insecticides might not be as straightforward as for forage crops. Herbicide and insecticide residues can present a very real and serious problem for livestock producers. This does not need to be a problem if the correct compounds are used and if they are used in strict accordance with label directions.

Sodium hydroxide, calcium hydroxide, and ammonia do not leave any toxic residues in treated feedstuffs. If these chemicals are sold for the purpose of increasing the digestibility of feedstuffs, U.S. Food and Drug Administration approval is probably necessary. Efficacy rather than safety is the primary concern.

RESEARCH NEEDS

Considerable research has been conducted on crop residues, but much remains to be done. The nutritional value of residues must be improved in many instances, and systems of collecting, storing, treating, and feeding must be improved to make increased use economical.

Nutritional improvement seems to be possible by plant genetic improvement, harvest management, chemical treatment, physical treatment, or biological treatment. All of these need further innovative research.

SUMMARY

The potential for increased use of crop residues in ruminant production is quite impressive, but the extent of usage depends on two important factors: the availability of grain for feeding to ruminants and the amount of research progress made in improving the nutritional value of crop residues.

Residues from the various crops vary in nutritional value, physical characteristics, and response to treatment. It is risky to generalize from one residue to another. Because of the low nutritional value of residues, methods of increasing value are of interest. Calcium hydroxide and ammonia treatments have the greatest long-term potential.

Livestock production systems can be developed to use large quantities of residues. These systems would not presently replace all of the grains, but would reduce the quantity used.

LITERATURE CITED

Acock, C. W. 1978. Wheat straw and sodium hydroxide treatment for beef cow maintenance diets. M.S. thesis. Lincoln: University of Nebraska.

Acock, C. W., J. K. Ward, I. G. Rush, and T. J. Klopfenstein. 1979. Wheat straw and sodium hydroxide treatment in beef cow rations. J. Anim. Sci. 49:354.

Anderson, D. C. 1978. Use of cereal straws in beef cattle production systems. J. Anim. Sci. 46:849.

Anderson, D. C., and A. T. Ralston. 1973. Chemical treatment of rye-grass straw: In vitro dry matter digestibility and compositional changes. J. Anim. Sci. 37:148.

Andreis, H. J., and R. P. DeStefano. 1978. Silage made from sugarcane bagasse treated with sodium hydroxide. P. 13 in Sugar J. October, 1978.

Asadpour, P. 1978. Utilization of treated wheat straw by sheep. M.S. thesis. Lincoln: University of Nebraska.

Barton, F. E., H. Amos, W. J. Albrecht, and D. Burdick. 1974. Treating peanut hulls to improve digestibility for ruminants. J. Anim. Sci. 38:860.

Bender, F., D. P. Heaney, and A. Bowden. 1970. Potential of steamed wood as feed for ruminants. For. Prod. J. 20:36.

Berger, L. L., J. A. Paterson, T. J. Klopfenstein, and R A. Britton. 1979. Effect of harvest data and sodium hydroxide treatment on the feeding value of corn stalkage. J. Anim. Sci. 49:1312.

Berger, L. L., T. J. Klopfenstein, and R. A. Britton. 1980. Rate of passage and rate of ruminal fiber digestion as affected by level of NaOH treament. J. Anim. Sci. 50:745.

Bolsen, K. K., C. Grimes, J. G. Riley, and L. Corak. 1976. Milo stover, forage sorghum and alfalfa silages for growing calves. P. 203 in ASAS Annual Meeting, St. Louis. (Abstr.)

Bolsen, K. K., C. Grimes, and J. G. Riley. 1977. Milo stover in rations for growing heifers and lambs. J. Anim. Sci. 45:377.

Bottje, W. G., B. L. Miller, L. L. Berger, R. B. Rindsig, and G. C. Fahey. 1979. In vivo and in vitro evaluations of soybean residues ensiled with various additives. P. 93 in ASAS Midwestern Section Meeting, St. Louis. (Abstr.)

Brown, W. H., F. M. Whiting, B. S. Daboll, R. J. Turner, and J. D. Schuh. 1977. Pelleted and nonpelleted cottonseed hulls for lactating dairy cows. J. Dairy Sci. 60:919.

Brown, W. H., G. D. Halbach, J. W. Stull, and F. M. Whiting. 1979. Utilization of cotton gin trash by lactating dairy cows. J. Dairy Sci. 62:793.

Burdick, D., F. E. Barton, and H. E. Amos. 1975. Performance of steers fed peanut hulls as roughages. Weight gains and DDT residues. ARS-S-61. New Orleans: U.S. Department of Agriculture.

Butterworth, M. H. 1962. The digestibility of sugar-cane tops, rice aftermath, and bamboo grass. Emp. J. Exp. Agric. 30:77.

Calhoun, M. C., and M. Shelton. 1973. Peanut Hulls and Cottonseed Hulls Compared with Alfalfa Hay as Roughage Sources in High Concentrate Lamb Rations. Tex. Agric. Exp. Stn. Rep. PR-3179.

Chandra, S., and M. G. Jackson. 1971. A study of various chemical treatments to remove lignin from coarse roughages and increase their digestibility. J. Agric. Sci. 77:11.

Choung, C. C., and W. R. McManus. 1976. Studies on forage cell walls. 3. Effects of feeding alkali-treated rice hulls to sheep. J. Agric. Sci. (Cambridge). 86:517.

Clanton, D. C., and L. Harris. 1966. Beet tops: Silage vs. pasturing. P. 12 in Nebraska Beef Cattle Progress Report. Lincoln: University of Nebraska.

Colenbrander, V. F., V. L. Lecktenberg, and L. G. Bauman. 1973. Digestibility of feeding value of brown midrib corn stover silage. J. Anim. Sci. 37:294.

Coombe, J. B., D. A. Dinius, H. K. Goering, and R. R. Oltjen. 1979a. Wheat straw urea diets for beef steers: Alkali treatment and supplementation with protein, monensin and a feed stimulant. J. Anim. Sci. 48:1223.

Coombe, J. B., D. A. Dinius, and W. E. Wheeler. 1979b. Effect of alkali treatment on intake and digestion of barley straw by beef steers. J. Anim. Sci. 49:169.

Corken, R., R. McElroy, H. Taylor, and W. B. Black. 1979. Feasibility and effects of increased use of crop residues in beef cattle rations. Washington, D.C.: Economics, Statistics and Cooperative Service, USDA.

Daniels, L. B., and R. B. Hashim. 1977. Evaluation of fungal cellulases in rice hull base diets for ruminants. J. Dairy Sci. 60:1563.

Dinusson, W. E. 1969. Straw for Wintering Cows. 20th Annual Research Roundup. Dickinson, North Dakota: Dickinson Exp. Stn.

Donefer, E., I. O. A. Adeleye, and T. A. O. C. Jones. 1969. Effect of urea supplementation on the nutritive value of NaOH-treated oats straw. Pp. 328-339 in Celluloses and Their Applications. Advances in Chemistry Series 95. Washington, D.C.: American Chemical Society.

Dronawat, N. W., R. W. Stanley, E. Cobb, and K. Morita. 1966. Effect of feeding limited roughage and a comparison between loose and pelleted pineapple hay on milk production, milk constituents, and fatty acid composition of milk fat. J. Dairy Sci. 49:28.

Durham, R. R., and D. D. Hinman. 1979. Digestibility and utilization of bluegrass straw harvested on three different dates. J. Anim. Sci. 48:464.

Early, R. J., and D. C. Anderson. 1978. Kentucky bluegrass straw composition, digestibility and utilization in wintering cow rations. J. Anim. Sci. 46:787.

Estima, A. L., G. C. Caldas, S. P. Viana, M. F. Cavalcante, A. R. de Carvalho, M. S. Farias, and G. P. Lofgreen. 1967. Molasses, cassava and cottonseed meal as supplements to fresh and ensiled sugarcane tops. IRI Res. Inst. Bull. 32. New York: IRI Research Institute.

Faulkner, D. G., and J. K. Ward. 1981. Cornstalk grazing of weanling heifers. Nebraska Beef Cattle Report. EC 80-218. Lincoln: University of Nebraska.

Faulkner, D. B., J. K. Ward, T. J. Klopfenstein, and I. B. Rush. 1981. Ammonia treatment of wheat straw for cows. Nebraska Beef Cattle Report. EC 80-218. Lincoln: University of Nebraska.

Garlich, J. D., D. M. Bryant, H. M. Covington, D. S. Chamblee, and A. E. Purcell. 1974. Egg yolk and broiler skin pigmentation with sweet potato vine meal. Poult. Sci. 53:692.

Garrett, W. N., H. G. Walker, G. O. Kohler, and M. R. Hart. 1979. Response of ruminants to diets containing sodium hydroxide on ammonia treated rice straw. J. Anim. Sci. 48:92.

Guggolz, J., G. O. Kohler, and T. J. Klopfenstein. 1971. Composition and improvement of grass straw for ruminant nutrition. J. Anim. Sci. 33:151.

Gupta, B. S., D. E. Johnson, and F. C. Hinds. 1978. Soybean straw intake and nutrient digestibility by sheep. J. Anim. Sci. 46:1086.

Haines, C. E., H. L. Chapman, and R. W. Kidder. 1959. The Feeding Value and Digestibility of Dried Celery Tops for Steers. Everglades Stn. Mimeo. Rep. 59-13. Gainesville: University of Florida.

Hale, W. H., C. Lambeth, B. Theurer, and D. E. Ray. 1969. Digestibility and utilization of cottonseed hulls by cattle. J. Anim. Sci. 29:773.

Hasimoglu, S., T. J. Klopfenstein, and T. H. Doane. 1969. Nitrogen source with sodium hydroxide treated wheat straw. J. Anim. Sci. 29:160.

Heinemann, W. W. 1976. Bluegrass Straw, Cottonseed Hulls and Wheat Screenings in Steer Finishing Rations. College of Agriculture, Washington State Univ. Bull. 832. Pullman: Washington State University.

Holloway, J. W., J. M. Anderson, W. A. Pund, W. D. Robbins, and R. W. Rogers. 1974. Feeding Gin Trash to Beef Cattle. Miss. Agric. For. Exp. Stn. Bull. 818.

Horton, G. M. J., and G. M. Steacy. 1979. Effect of anhydrous ammonia treatment on the intake and digestibility of cereal straws by steers. J. Anim. Sci. 48:1239.

Hunt, G. C., K. R. Cummings, and J. W. Lusk. 1971. Cottonseed hull-concentrate complete rations for lactating cows. J. Dairy Sci. 54:452. (Abstr.)

Hutanuwatr, N., F. C. Hinds, and C. L. Davis. 1974. An evaluation of methods for improving the in vitro digestibility of rice hulls. J. Anim. Sci. 38:140.

Jackson, M. G. 1977. Review article: The alkali treatment of straws. Anim. Feed Sci. Technol. 2:105.

Jackson, M. G. 1978. Treating Wheat Straw for Animal Feeding. FAO Animal Production and Health Paper No. 10. Rome: Food and Agriculture Organization of the United Nations.

Jared, A. H., and E. Donefer. 1970. Alkali-treated straw rations for fattening lambs. J. Anim. Sci. 31:245.

Jayasuriya, M. C. N., and E. Owen. 1975. Sodium hydroxide treatment of barley straw: Effect of volume and concentration of solution on digestibility and intake by sheep. Anim. Prod. 21:313.

Jones, J. H., D. S. Logan, and P. J. Lyerly. 1957. Use of Cotton Gin Trash in Steers' Fattening Rations. Tex. Agric. Exp. Stn. Prog. Rep. 1969.

Karn, H. P., and K. S. Hendrix. 1977. Digestibility of alkali-treated and brown corn plant residue. P. 611 in ASAS Annual Meeting. Madison: University of Wisconsin. (Abstr.)

Kellems, R. O., O. Wayman, A. H. Nguyen, J. C. Nolan, C. M. Campbell, J. R. Carpenter, and E. B. Ho-a. 1979. Post-harvest pineapple plant forage as a potential feedstuff for beef cattle: Evaluated by laboratory analyses, in vitro and in vivo digestibility and feedlot trials. J. Anim. Sci. 48:1040.

Kernan, J. A., E. C. Coxworth, and M. J. Moody. 1980. A survey of the feed value of various specialty crop residues and forages before and after chemical processing. Saskatchewan Research Council Publication No. C-814-F-1-B-80. Saskatoon: University of Saskatchewan.

Klopfenstein, T. J. 1975. Pressure treatment of corn cobs. In Nebraska Beef Cattle Report. EC 75-218. Lincoln: University of Nebraska.

Klopfenstein, T. J. 1978. Chemical treatment of crop residues. J. Anim. Sci. 46:841.

Klopfenstein, T. J., and K. K. Bolsen. 1971. High temperature pressure treated crop residues. J. Anim. Sci. 33:290.

Klopfenstein, T. J., V. E. Krause, M. J. Jones, and W. Woods. 1972. Chemical treatment of low quality roughages. J. Anim. Sci. 35:418.

Klopfenstein, T. J., R. P. Graham, H. G. Walker, and G. O. Kohler. 1974. Chemicals with pressure treated cobs. J. Anim. Sci. 39:243.

Koers, W., W. Woods, and T. J. Klopfenstein. 1970. Sodium hydroxide treatment of corn stover and cobs. J. Anim. Sci. 31:1030.

Koers, W., M. Prokop, and T. J. Klopfenstein. 1972. Sodium hydroxide treatment of crop residues. J. Anim. Sci. 35:1131.

Lamm, W. D. 1976. Influence of nitrogen supplementation on hydroxide treatment upon the utilization of corn crop residues by ruminants. Ph.D. dissertation. Lincoln: University of Nebraska.

Lamm, W. D., and J. K. Ward. 1977. Corn crop residue quality and compositional changes. In ASAS 69th Annual Meeting. Madison: University of Wisconsin. (Abstr. 46)

Larson, W. E., R. F. Holt, and C. W. Carlson. 1978. Residues for soil conservation. Chap. 1 in Crop Residue Management and System. Madison, Wisc.: American Society of Agronomy.

Lau, M. M., and P. J. Van Soest. 1981. Titratable groups and soluble phenolic compounds as indicators of the digestibility of chemically treated roughages. Anim. Feed Sci. Technol. 6:123–131.

Leonard, R. O., M. E. Stanley, and D. L. Bath. 1973. Unusual Feedstuffs in Livestock Rations. U. Calif. Agric. Ext. Bull. AXT-1979.

Lesoing, G., I. Rush, T. Klopfenstein, and J. Ward. 1980a. Wheat straw in growing cattle diets. J. Anim. Sci. 51:257.

Lesoing, G., T. Klopfenstein, I. Rush, and J. Ward. 1980b. Chemical treatment of wheat straw. J. Anim. Sci. 51:263.

Livingston, A. L., R. E. Knowles, J. Page, D. D. Kuzmicky, and G. O. Kohler. 1972. Processing of cauliflower leaf waste for poultry and animal feed. J. Agric. Food Chem. 20:277.

Marshall, S. P., and H. H. Van Horn. 1975. Complete rations for dairy cattle. II. Sugarcane bagasse pellets as roughage in blended rations for lactating cows. J. Dairy Sci. 58:896.

Martin, P. C., T. C. Cribeiro, A. Cabello, and A. Elias. 1974. The effect of sodium hydroxide and pressure on the dry matter digestibility of bagasse and bagasse pith. Cuban J. Agric. Sci. 8:21.

Martin, P. C., A. Cabello, and A. Elias. 1976. The use of fibrous sugarcane by-products by ruminants. 2. Effect of the NaOH pressure combination on the digestibility and chemical composition of bagasse and bagasse pith. Cuban J. Agric. Sci. 10:19.

McDonnell, M. L., and T. J. Klopfenstein. 1980. Cornstalk quality as affected by variety and management. In ASAS Annual Meeting. Ithaca, N.Y.: Cornell University. (Abstr.)

McKee, M., K. G. Kimple, and K. K. Bolsen. 1977. Crop residues for wintering beef cows in dry lot. In ASAS Annual Meeting. Madison: University of Wisconsin. (Abstr. 314)

McManus, W. R., and C. C. Choung. 1976. Studies on forage cell walls. 2. Conditions for alkali treatment of rice straw and rice hulls. J. Agric. Sci. (Cambridge) 86:453.

National Research Council. 1971. Atlas of Nutritional Data on United States and Canadian Feeds. Washington, D.C.: National Academy of Sciences.

Noland, P. R., and B. F. Ford. 1954. Rice hulls and rice mill feed. Ark. Farm Res. 3(3).

Ololade, B. G., D. N. Mowat, and J. E. Winch. 1970. Effect of processing methods on the in vitro digestibility of sodium hydroxide treated roughages. Can. J. Anim. Sci. 50:657.

Otagaki, K. K., G. P. Lofgreen, E. Cobb, and G. G. Dull. 1961. Net energy of pineapple bran and pineapple hay when fed to lactating dairy cows. J. Dairy Sci. 44:491.

Pate, F. M., and S. W. Coleman. 1975. Sugarcane Tops for Cattle Feed. Fla. Agric. Exp. Stn. J. Series No. 5509.

Pate, F. M., D. W. Beardsley, and B. W. Hays. 1971. Chopped Sugarcane Tops as a Feedstuff for Cattle and Horses. Everglades Exp. Stn. Mimeo. Rep. EES71-5.

Paterson, J. A. 1979. The feeding of hydroxide treated crop residues to growing ruminants. Ph.D. thesis. Lincoln: University of Nebraska.

Paterson, J. A., T. J. Klopfenstein, and R. A. Britton. 1978a. The digestibility of sodium hydroxide treatments of roughage on mineral balance and digestibility. J. Anim. Sci. 46:340. (Abstr.)

Paterson, J. A., T. J. Klopfenstein, and R. A. Britton. 1978b. Effect of sodium hydroxide treatments of roughage on mineral balance and digestibility. J. Anim. Sci. 46:433. (Abstr.)

Paterson, J. A., M. L. McDonnell, and T. J. Klopfenstein. 1979. Decrease in digestibility of cornstalks after physiological maturity. P. 66 in ASAS Midwest Section Meeting. (Abstr.)

Paterson, J. A., R. Stock, and T. J. Klopfenstein. 1980. Calcium hydroxide treatment of crop residues. Nebraska Beef Cattle Report. EC 80-218:21. Lincoln: University of Nebraska.

Randel, P. F. 1970a. Dairy beef production from mixtures of sugarcane bagasse and concentrates. J. Agr. Univ. P. R. 54:237.

Randel, P. F. 1970b. Ad libitum feeding of either a complete ration based on sugarcane bagasse or a conventional concentrates mixture to dairy cows. J. Agr. Univ. P. R. 54:429.

Randel, P. F. 1972. A comparison of the digestibility of two complete rations containing either raw or alkali-treated sugarcane bagasse. J. Agr. Univ. P. R. 56:18.

Randel, P. F., A. Ramirez, R. Carrero, and I. Valencia. 1971. Alkali-treated and raw sugarcane bagasse as roughages in complete rations for lactating cows. J. Dairy Sci. 55:1492.

Rexen, B. 1977. Enzyme solubility—A method for evaluating the digestibility of alkali treated straw. Anim. Feed Sci. Technol. 2:205.

Rexen, R., and K. V. Thomsen. 1976. The effect on digestibility of a new technique for alkali treatment of straw. Anim. Feed Sci. Technol. 1:73.

Rogers, D. J., and M. Milner. 1963. Amino acid profile of manioc leaf protein in relation to nutritive value. Econ. Bot. 17:211.

Ross, E., and F. Q. Enriquez. 1969. The nutritive value of cassava leaf meal. Poult. Sci. 48:846.

Rush, I. 1977. Sugar beet by-products—What are they worth? P.481 in Beef Cattle Science Handbook 14. Clovis, Calif.: Agriservices Foundation.

Saxena, S. K., D. W. Otterby, J. D. Donker, and A. L. Good. 1971. Effects of feeding alkali-treated oat straw supplemented with soybean meal or non-protein nitrogen on growth of lambs and on certain blood and rumen liquor parameters. J. Anim. Sci. 33:485.

Sherrod, L. B., and C. B. Summers. 1974. Sodium hydroxide treatment of cottonseed hulls and sorghum stubble. In ASAS Proceedings, Western Section. Corvallis, Ore. (Abstr. 388)

Singh, M., and M. G. Jackson. 1971. The effect of different levels of sodium hydroxide spray treatment of wheat straw on consumption and digestibility by cattle. J. Agric. Sci. 77:5.

Smith, D. H. 1977. Effect of physiological and management factors on yield and quality of grain sorghum residues. Ph.D. thesis. Lincoln: University of Nebraska.

Sundstol, F., E. Coxworth, and D. N. Mowat. 1978. Improving the nutritive value of straw and other low-quality roughages by treatment with ammonia. World Anim. Rev. 26:13.

Tillman, A. D., R. D. Furr, D. R. Hansen, L. B. Sherrod, and J. D. Word. 1969. Utilization of rice hulls in cattle finishing rations. J. Anim. Sci. 29:792.

U.S. Department of Agriculture. 1979. Agricultural Statistics. Washington, D.C.: U.S. Government Printing Office.

Umunna, N. N., and T. J. Klopfenstein. 1972. Response of lambs fed pressure treated wheat straw. J. Anim. Sci. 35:1136.

Umunna, N. N., T. J. Klopfenstein, and K. Bolsen. 1972. Response of lambs fed pressure treated corn cobs. J. Anim. Sci. 35:277.

Utley, P. R., and W. C. McCormick. 1972. Level of peanut hulls as a roughage source in beef cattle finishing diets. J. Anim. Sci. 34:146.

Van Soest, P. J. 1981. Limiting factors in plant residues of low biodegradability. Agric. Environ. 6:135–143.

Vetter, R. L. 1973. Evaluation of chemical and nutritional properties of crop residues. In Crop Residue Symposium. Lincoln: University of Nebraska.

Vetter, R. L., and M. Boehlje. 1978. Alternative feed resources for animal production. In Plant and Animal Products in the U.S. Food System. Washington, D.C.: National Academy of Sciences.

Waagepetersen, J., and K. V. Thomsen. 1977. Effect on digestibility and nitrogen content of barley straw of different ammonia treatments. Anim. Feed Sci. Technol. 2:131.

Waiss, A. C., Jr., J. Guggolz, G. O. Kohler, H. G. Walker, Jr., and W. N. Garrett. 1972. Improving digestibility of straws for ruminant feed by aqueous ammonia. J. Anim. Sci. 35:109.

Waller, J. C. 1976. Evaluation of sodium, calcium, and ammonium hydroxides for treating residues. M.S. thesis. Lincoln: University of Nebraska.

Waller, J. C., and T. J. Klopfenstein. 1975. Hydroxides for treating crop residues. J. Anim. Sci. 41:424.

Ward, J. K. 1978. Utilization of corn and grain sorghum residues in beef cow forage systems. J. Anim. Sci. 46:831.

Ward, J. K., L. J. Perry, D. H. Smith, and J. T. Schmitz. 1979. Forage composition and utilization of grain sorghum residue by beef cows. J. Anim. Sci. 48:919.

White, T. W. 1966. Utilization of ammoniated rice hulls by beef cattle. J. Anim. Sci. 25:25.

Wilkinson, J. M., and R. Gonzalez Santillana. 1978. Ensiled alkali-treated straw. I. Effect of level and type of alkali on the composition and digestibility in vitro of ensiled barley straw. Anim. Feed Sci. Technol. 3:117.

GLOSSARY TERMS

BAGASSE	Solid residue remaining after extraction of juice.
CHAFF	Husks, hulls, joints, and small fragments of straw that are separated from seed in threshing of small grains.
CORNCOBS	Fibrous portion of the fruiting head (grain producing portion) of the corn plant, excluding the husk.
CORN HUSK	Fibrous outside cover of the fruiting head (grain producing portion) of the corn plant.
CROP RESIDUE	Fibrous residue remaining after harvest of the primary product (grain, fruit, etc.).
HULLS	Outer covering of seeds.
HUSK	Outer covering of kernels or seeds, especially when fibrous.
HUSKLAGE	Corncobs and husks.
IN VITRO	Outside the living organism in an artificial environment.
IN VIVO	Within the living organism.
POD	Empty seed vessel.
STALK	Main stem of a herbaceous plant.
STALKLAGE	Moist, ensiled stalks.
STOVER	Stalks and leaves of corn or sorghum after grain harvest.
STRAW	Plant residue remaining after separation of the seeds by threshing of small grains.
STUBBLE	Lower parts of plant stems that remain standing in the field after harvest.

6
Aquatic Plants

INTRODUCTION

Aquatic plants occur throughout the world in oceans, saltwater marshes, rivers, lakes, and waste-treatment ponds. Seaweeds, especially kelp, have been used as medicinals for centuries, but little use has been made of other aquatic plants.

These plants have been regarded more as problems than resources. A National Academy of Sciences report pointed out that the problem of aquatic weeds was reaching alarming proportions in many parts of the world (National Research Council, 1976). The report pointed out the following adverse effects of these plants: blocking canals and pumps in irrigation projects, interfering with hydroelectric production, wasting water by evapotranspiration, hindering boat traffic, increasing waterborne disease, interfering with fish culture and fishing, and impeding drainage, which results in flooding. The problem seems to be more severe in tropical areas. Aquatic plants may foster mosquitoborne diseases because small sheltered pools formed between floating plants are well adapted for mosquito breeding.

Recently, aquatic plants have been recognized as potentially valuable resources for animal feed and other uses. The production per hectare may be quite large. Problems exist in harvesting and processing these plants for animal feed, since they are grown in water and are very high in moisture content. Nevertheless, they appear to be valuable resources for use as feedstuffs for production of meat, milk, and eggs.

QUANTITY

Information is available on seaweed production by countries known to harvest seaweeds (Naylor, 1976). The values by countries for 1960, 1967, and 1973 are given in Table 59. The total tonnage harvested is not very large, especially considering the low dry-matter content of the harvested seaweed.

Data are not available on total quantity of other aquatic plants; however, there is potential for very large quantities. For example, growth rates of 800 kg dry matter/ha/day have been recorded for water hyacinth (National Research Council, 1976). If such growth could be expected for even 6 months a year, a yield of 146 tons dry matter/ha would be achieved. Yields of 17.8 tons dry matter/ha were reported for duckweeds, compared to 4.4 to 15.9 tons for alfalfa for hay (Hillman and Culley, 1978). Standing crops of microalgae have yielded up to 1,130 g dry weight/m^2 surface pond (Boyd, 1973a). This would amount to 11.3 tons dry matter/ha.

PHYSICAL CHARACTERISTICS

The physical characteristics of aquatic plants present problems in harvesting and processing. Algae are small (5 to 15 μm) and have low specific gravity (Golueke and Oswald, 1965). Kelp varies in length, from very small to 6 to 9 m long (Hart et al., 1978). Water hyacinths consist of a short rhizome, roots, rosulate leaves, and inflorescence and stolons that connect different plants (Pieterse, 1978). Duckweed is a tiny free-floating vascular plant (Rusoff et al., 1980). Harvested plants usually consist of a slippery, tangled mass that is difficult to handle mechanically. Water hyacinths grow rapidly and may present problems by blocking water flow, thus interfering with rice production. In Bangladesh, rafts of water hyacinths weighing up to 300 tons/ha float over rice paddies (National Research Council, 1976).

A common physical characteristic of all water plants is a low dry-matter content, varying from 5 to 15 percent (National Research Council, 1976). Kelp contains approximately 12 percent dry matter (Hart et al., 1978). The low dry-matter content of algae in ponds is an even more serious problem; algae harvested from ponds contains only 1 to 2 percent solids (Golueke and Oswald, 1965).

TABLE 59 Production of Seaweeds and Aquatic Plants

Continent	Country	Thousands Metric Tons (wet weight) 1960	1967	1973
Africa	Egypt	—	—	3.8
	Morocco	17.0	18.0	8.0
	Senegal	—	—	(1.0)
	South Africa	—	43.0	24.1
	Tanzania	0.5	2.0	0.5
North and Central America	Canada	13.2	44.1	39.9
	Mexico	15.5	28.9	36.9
	United States	(91.0)	2.3	1.1
South America	Argentina	0.9	32.8	24.4
	Brazil	—	—	(103.0)
	Chile	7.2	32.5	26.5
	Peru	—	(1.0)	(1.0)
Asia	China	(250.0)	—	(700.0)
	India	(1.5)	(3.5)	(5.5)
	Indonesia	—	—	16.6
	Japan	386.1	534.3	654.2
	Republic of Korea	29.7	87.8	224.2
	Philippines	(0.1)	1.6	2.8
	Thailand	—	0.3	—
Europe	Denmark	15.6	27.5	10.9
	France	45.7	76.3	60.2
	Iceland	(20.0)	—	(20.0)
	Ireland	69.1	(54.0)	(44.0)
	Italy	—	—	(0.6)
	Norway	—	61.0	(75.0)
	Spain	15.5	36.0	47.0
	United Kingdom	(18.0)	23.1	24.1
	USSR	—	—	(100.0)
Oceania	Australia	—	(7.0)	—
	New Zealand	(0.4)	(0.5)	(0.6)
World Total		1171	1886	2402

NOTE: Numbers in parentheses are estimates.

SOURCE: Naylor (1976). Courtesy of the Food and Agriculture Organization of the United Nations.

NUTRITIVE VALUE

Chemical Composition

Algae

Crude protein content of algae (see Appendix Table 1) ranges from 31 to 68 percent, dry-matter basis. The levels of essential amino acids indicate that algae protein is of high quality (see Appendix Table 4). Algae are also rather high in ether extract and low in crude fiber, indicating fair energy value. Values given for required minerals indicate that algae are good sources of these (see Appendix Table 2).

Other Aquatic Plants

The crude protein content of other aquatic plants ranges from 11 to 32 percent, dry basis (see Appendix Table 1). These levels would be satisfactory for ruminants and for most classes of swine if the amino acid pattern is satisfactory. Aquatic leaf protein was lower in lysine and methionine than meat proteins (Boyd, 1968). The fiber level appears to be lower in duckweed than in other aquatic plants. Aquatic plants are also rich in required minerals (see Appendix Table 2). According to Bagnall et al. (1973) 3 kg dry water hyacinths would supply an excess of the major minerals for 450 kg steers. None of the other minerals were present at high levels (see Appendix Table 3).

Nutrient Utilization

Algae

The protein efficiency ratio (PER) in rats was lower for dried algae than for casein (Cook, 1962). Autoclaving algae lowered the PER, and cooking for 30 minutes or 2 hours had no effect. However, cooking improved true digestibility of protein and net protein utilization (NPU). The PER of protein in three species of algae was lower than that of soy protein and casein for chicks and rats, respectively (McDowell and Leveille, 1963). Supplementing algae protein with methionine increased the PER for rats but not to the value for casein. Including cellulase, diastase, and alpha amylase at 1 percent of the diet increased the digestibility of nitrogen by laboratory rats. In swine, digestibility of the protein in algae was 72 percent, whereas the digestibility of energy was only 34 percent (Hintz and Heitman, 1967). PER, biological value, and apparent digestibility of nitrogen were lower for algae protein than for casein in growing rats (Chung et al., 1978).

Dry-matter digestibility by sheep of dried or dehydrated algae was 54.2 percent, calculated by difference (Hintz et al., 1966). Energy digestibility was 58.7 percent. In cattle, energy digestibility of algae was 59 percent.

Seaweed

Incorporation of 20 percent seaweed in a concentrate mixture did not affect digestibility of protein, but lowered the TDN (67.9 versus 70.6 percent) of a calf diet (Shukla et al., 1974). Digestion coefficients in pigs were 12 percent for energy and negative for nitrogen in seaweed residue that had been extracted for alginate (Whittemore and Percival, 1975). The authors indicated that the nitrogen fraction was largely insoluble. They indicated that the low energy digestibility resulted from a loss of soluble energy through the extraction process.

Water Hyacinth

Organic matter and crude protein digestibility were lower for ensiled water hyacinth and dried citrus pulp (4 percent) than for pangolagrass and citrus pulp (Bagnall et al., 1974). Values for organic-matter digestibility were 40 and 48 percent for water hyacinth silages and 54 percent for pangolagrass. Crude protein digestibility averaged 51 percent for ensiled water hyacinth-dried citrus pulp for hyacinth from oxidation pond water, compared with 53 percent for hyacinth from lake water. Addition of 2 percent acetic acid, propionic acid, or a combination of these prior to ensiling was necessary to get a pH below 5 from ensiling partially dried aquatic plants (Linn et al., 1975a). Adding 5 percent corn grain did not lower the pH to less than 5. Dry-matter digestibilities were 43.8 and 43.4 percent for dried aquatic plants (two species), compared with 50.8 percent for dehydrated alfalfa (Linn et al., 1975b). Respective values for energy digestibility were 53.7, 47.4, and 49.2 percent. Digestibility of ensiled aquatic plants was 41.4 percent for dry matter and 51.4 percent for energy. Crude protein digestibility was 58.3 percent for dried material and 33.0 percent for the ensiled material.

Replacing 35 percent of alfalfa meal with dried aquatic plants depressed dry-matter digestibility from 77 to 72 percent in sheep and 66 to 64 percent in goats (Heffron et al., 1977). In vivo digestibilities of dry matter, organic matter, and crude protein were lower for ensiled water hyacinth than ensiled pangolagrass (Baldwin et al., 1975). Values for pangolagrass and water hyacinth, respectively, were 54.0 and 35.0 percent for dry matter,

53.6 and 39.6 percent for organic matter, and 76.1 and 52.8 percent for crude protein. Digestibility of amino acids in duckweed by channel catfish varied from 38 percent for glycine to 64 percent for tyrosine (Robinette et al., 1980).

Animal Performance

Algae

Chicks could tolerate only up to 10 percent algae meal in the diet if residual alum was present, but could tolerate up to 20 percent of aluminum-free algae meal (Grau and Klein, 1957). Algae was found to contain substantial amounts of xanthophyll, which can be utilized in the chick for shank and skin pigmentation. Performance of chicks and rats fed algae-supplemented diets was lower than that of comparable animals fed soybean-protein and casein-supplemented diets (McDowell and Leveille, 1963). Supplementation of algae-containing diets with seven essential amino acids increased growth rate 49 percent and feed efficiency 47 percent in chicks, indicating amino acid deficiency in algae protein.

Including up to 10 percent algae in swine diets did not alter rate of gain or feed efficiency (Hintz et al., 1966). Supplementing lambs on summer range in California with pellets in which 50 percent of the nitrogen was supplied by algae and 50 percent by alfalfa resulted in higher weight gains than supplementing only with alfalfa pellets. Rate of gain was similar for pigs fed diets supplemented with algae or fishmeal if certain B-vitamins were also supplemented (Hintz and Heitman, 1967). Performance was lowered if the diets were supplemented with only algae. In a subsequent trial, evidence was obtained that the effect of B-vitamin supplementation was due at least partly to vitamin B_{12}.

Seaweed

Including 20 percent seaweeds in the concentrate of a calf diet tended to lower rate of gain, but the difference was not significant (Shukla et al., 1974). Feeding concentrate mixtures with up to 30 percent seaweed did not alter milk yields and fat-corrected milk yields in dairy cows (Desai and Shukla, 1975). Body weights tended to be lower if the level of seaweed meal exceeded 15 percent in the concentrate. Including up to 15 percent dried seaweed meal did not affect feed intake and weight gain in sheep (Herbert and Wiener, 1978). Diarrhea was observed in swine fed a diet containing 50 percent dried residue of seaweed after extraction of alginate (Whittemore and Percival, 1975).

Water Hyacinth

Body weights and shank pigmentation score of chicks fed a diet containing duckweed were similar to those of chicks fed diets containing alfalfa (Truax et al., 1972).

Cattle preferred water hyacinth silage containing the highest level of preservative, 4 kg dried citrus pulp and 1 kg sugarcane molasses/100 kg plant residue (Baldwin et al., 1974). The silage preference was negatively related to pH and ash in silage. Cattle and sheep readily consumed complete diets with processed water hyacinth (Bagnall et al., 1974). Dry-matter intake of pangolagrass silage by sheep was higher than that of water hyacinth silage (Baldwin et al., 1975). Intake of a diet by sheep and goats in which dried aquatic plant was substituted for 35 percent of the alfalfa meal was similar to that of the control diet (Heffron et al., 1977).

Rate of gain of dairy heifers fed diets containing duckweed was equal to or higher than that of heifers fed control diets (Rusoff et al., 1977).

PROCESSING

Harvesting and processing water plants would result in reduction of weeds in waterways in addition to providing a resource that could be used as animal feed. There are problems with harvesting and processing. Since weeds are located in water, specially adapted harvesting equipment is essential. The high water content (up to 95 percent) adds volume and weight to the process and means some of the water may have to be removed.

Algae

Harvesting algae involves three steps: (1) initial concentration, (2) dewatering, and (3) final drying (Golueke and Oswald, 1965). The solids content of pond water is increased to 1 to 2 percent in the initial concentration. Chemical precipitation and centrifuging appear to be most promising for the initial concentration. Suitable floculating agents were alum, lime, and synthetic organic polyvalent cationic polymers. Optimum pH for alum precipitation was 6 to 7. Optimum pH for precipitation with lime was 11.2.

Excellent dewatering was achieved by using a modified industrial gravity filter with two batch-type centrifuges and with a continuous solid-bowl centrifuge. In the gravity filtration, at maximum production of slurry of 14 m^3/ha/day, a bed width of 24 m would be required to process the yield from each 0.4 ha of pond.

Golueke and Oswald (1965) suggested minimum exposure to high temperatures during drying in order to maintain quality. They suggested flash drying, a process similar to alfalfa dehydration procedures. A high-grade product was obtained using a steam-heated drum dryer. These workers used an inexpensive method of sand bed dewatering and drying of algae. An area of 1,470 m^2 would be required/ha of pond.

Seaweed

The greatest problem facing the seaweed industry is the need for improved efficiency in harvesting (Naylor, 1976). Hand collection still accounts for most seaweed harvested. Few mechanized methods have been developed successfully, the principal exception being the California *Macrocystis* beds. Compared with the brown algae, red algae are much smaller and occur in deeper water, making their harvesting slower and more costly. Red algae are usually collected by hand or by using rakes or grapnels. Along the eastern seaboard of the United States and Canada, *Chondrus crispus* (Irish moss) is normally harvested by use of long-handled rakes from small boats.

After harvesting, the bulk of seaweed must be reduced by drying before further processing, since the wet material deteriorates in quality (Naylor, 1976). Most seaweed is dried naturally by exposure to sun and air, but there has been increased use of mechanical dehydration. In the case of natural drying, the washed seaweeds are spread thinly on portable racks, wooden platforms, or areas of flat rock. Under favorable drying conditions 1 or 2 days are sufficient to reduce the moisture level to 18 to 20 percent.

The problems and uncertainties of the weather have stimulated the development of artificial drying, especially the use of rotary flame-heated air driers. The use of this method of drying also results in a more uniform product.

A process has been developed to reduce water and ash content of kelp and increase its caloric and bulk densities (Hart et al., 1978). The process consists of chopping the kelp in a silage cutter, then grinding in a vertical-shaft hammermill. After addition of calcium chloride the mixture is heated and pressed, with removal of 75 percent of the moisture and 65 percent of the ash. The level of calcium chloride was 0.5 percent. Following calcium chloride treatment, kelp could be pressed successfully after 15 minutes at 90°C or 30 minutes at 77°C. The process is shown diagramatically in Figure 14.

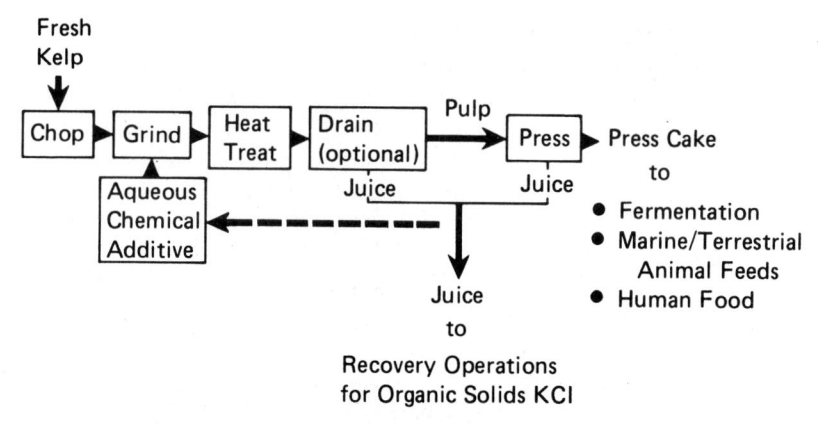

FIGURE 14 Flow diagram of kelp dewatering process. SOURCE: Hart et al. (1978).

Other Aquatic Plants

Aquatic plants may be harvested from a site at the water's edge or with a self-propelled floating harvester (National Research Council, 1976). Small moving boats can be used to mow aquatic plants (see Figure 15). A harvester for submerged plants is shown in Figure 16. Stationary conveyers (see Figure 17) may be used to remove aquatic plants from the water and transport them to processing or transportation equipment (Bryant, 1973).

Approximately 50 percent of the water in aquatic plants can be removed simply by pressing since it is on the surface or contained loosely in the vascular system (National Research Council, 1976). This press water contains only about 2 percent of the plant solids. Heavier pressing is necessary for further reduction in water. Approximately 10 to 30 percent of the solids may be removed by further pressing. The complete process may remove up to 70 percent of the water in hyacinth, reducing the moisture level from about 95 to 65 percent. This moisture level would enable ensiling of the aquatic plant. Bagnall and Hentges (1979) produced cattle feed by harvesting, chopping, pressing, and ensiling water hyacinth. Production energy and cost were less and the ensiled product was more acceptable to cattle than dried feed. It appears that artificial drying of aquatic plants is not economically feasible because of the high-moisture content of these materials. In areas of low rainfall, solar drying may be feasible.

FIGURE 15 Harvester for water hyacinth. An experimental harvester lifts, breaks, and throws water hyacinth by grasping and pulling on aerial parts of the plants, protecting the high-speed harvesting mechanism from damage by sudden underwater obstructions. SOURCE: National Research Council (1976).

UTILIZATION SYSTEMS

The high protein and mineral levels in algae indicate that these materials could be dried and used in limited amounts in swine and poultry diets. They have been used at levels of 10 to 20 percent of the diet in these animals (Grau and Klein, 1957; Hintz et al., 1966). The lower protein and energy value of the other materials would make them more likely to be used for ruminants. Seaweeds are usually used as sources of minerals and vitamins rather than of protein and energy (Stephenson, 1973). It is probable that ensiling would be more economically feasible than dehydration.

ANIMAL AND HUMAN HEALTH

Feeding a diet containing 18.3 percent algae (*Arthrospira*) to male rats for 14 days did not produce gross pathological or histopathological effects in liver, kidney, heart, lungs, spleen, stomach, intestines, testes, or lymph nodes (Chung et al., 1978). Feeding this diet to female rats from weaning

FIGURE 16 Floating harvester for submerged aquatic plants. Large, complete floating harvesters for submerged aquatic plants are impressive but expensive and have modest capacity. SOURCE: National Research Council (1976).

until the second week of lactation did not alter reproduction and lactation. The livers of these females were histologically normal. Feeding high levels of water hyacinth (*Eichhornia crassipes*) for three generations affected reproduction in rats (Chavez et al., 1976). During the first two generations little difference was noted in number of dams that littered when fed diets containing 0, 10, 20, and 30 percent dehydrated water hyacinth. During the third generation 75 and 38 percent of the females littered on the 20 and 30 percent hyacinth diets, respectively. The number of offspring weaned in the first generation was lower for females fed 30 percent hyacinth and was lower for females fed 20 and 30 percent hyacinths during the second or third generations. In all three generations, gains of rats from 5 through 13 weeks of age were lowest for those fed 30 percent hyacinth. The gains for those fed 10 and 20 percent hyacinth were lower than for the controls during the second and third generations. The nitrate, oxalate, and tannin content of hydrilla and water hyacinth did not appear to be a hazard in cattle diets (Stephens et al., 1973).

Data are not available concerning pathogens in water plants. The main potential problem appears to be in plants harvested in waste-treatment

FIGURE 17 Hyacinth control station. SOURCE: Bryant (1973). Courtesy of Aquamarine Corp. Crown Copyright, 1973, Controller of Her Britannic Majesty's Stationery Office.

ponds and lagoons. Data on mineral composition do not implicate a serious heavy metal threat (see Appendix Table 2), but this aspect needs to be explored further.

REGULATORY ASPECTS

There are no specific regulations concerning feeding water plants except those related to algae feeding. More data are needed before sound regulations can be formulated.

RESEARCH NEEDS

More critical experiments are needed to define the utilization of protein and energy in aquatic plants, especially seaweeds and water hyacinths.

This should include research on parts of plants, aerial versus submerged; stage of maturity; and palatability.

The most serious void is in development of economical mechanical harvesting and processing methods. The research should be with seaweed, algae, water hyacinths, and duckweed. In the case of algae, hyacinths, and duckweed, equipment needs to be developed for handling plants from lakes and waste-treatment ponds. Research should be directed toward developing controlled production, harvesting, processing, and quality for use as animal feed.

Research should be directed toward the health aspects of utilizing these resources. The presence of pathogens and methods of destroying these should be studied. Data should be collected on pesticides, mycotoxins, and heavy metals in these resources.

SUMMARY

Aquatic plants occur throughout the world and may present problems in irrigation projects, hydroelectric production, boat traffic, fish culture, and drainage. They are potentially valuable resources for different uses, including as animal feed. Yields of dry matter per hectare are usually higher for these plants than for conventional forage and grain crops. The physical characteristics, including varying size and high water contents, present problems in harvesting and processing the plants.

Generally, algae are high in crude protein and ether extract, and low in fiber. Water hyacinths are fairly high in fiber, indicating limited energy value, and are lower in protein than algae. Efficiency of utilization of protein in algae is good, although lower than in casein. The energy value of water hyacinths is lower and ash content much higher than for conventional roughages.

Including 10 percent algae in diets of chicks and swine did not alter performance. Including limited levels of seaweeds in ruminant diets has not altered performance. Water hyacinths have been ensiled successfully, and the silage was consumed readily by cattle and sheep.

The location of water plants and their high water content present problems in harvesting and processing. Equipment has been developed for processing various aquatic plants, but generally efficiency and capacity are low. Drying algae and ensiling the other aquatic plants probably are the most feasible processing methods at present.

Feeding 18 percent algae did not produce harmful effects in male and female rats. Feeding 10 percent dehydrated water hyacinths did not affect reproduction in female rats during three generations, but feeding 20 or 30 percent lowered reproductive rates. There is no evidence of harmful levels of minerals in aquatic plants.

LITERATURE CITED

Anonymous. 1970. State of development of the IFP Algae Process at December 1970. Report to FAO/WHO/UNICEF Protein Advisory Group. Institut Francais du Petrale.

Bagnall, L. O., and J. F. Hentges, Jr. 1979. Processing and conservation of water hyacinth and hydrilla for livestock feeding. In Aquatic Plants, Lake Management, and Ecosystem Consequences of Lake Harvesting. Proceedings of Conference at Madison, Wisconsin, Feb. 14–16, 1979. Madison: University of Wisconsin. 367 pp.

Bagnall, L. O., R. L. Shirley, and J. F. Hentges. 1973. Processing, chemical composition and nutritive value of aquatic weeds. P. 40 in Water Resources Research Center Pub. 25. Gainesville: University of Florida.

Bagnall, L. O., T. de S. Furman, J. F. Hentges, Jr., W. J. Nolan, and R. L. Shirley. 1974. Feed and fiber from effluent-grown water hyacinth. Pp. 116–141 in Wastewater Use in the Production of Food and Fiber. EPA Paper No. 660/2-74-041. Washington, D.C.: U.S. Government Printing Office.

Baldwin, J. A., J. F. Hentges, Jr., and L. O. Bagnall. 1974. Preservation and cattle acceptability of water hyacinth silage. Hyacinth Control J. 12:79–81.

Baldwin, J. A., J. F. Hentges, Jr., L. O. Bagnall, and R. L. Shirley. 1975. Comparison of pangolagrass and water hyacinth silages as diets for sheep. J. Anim. Sci. 40:968–971.

Boyd, C. E. 1968. Fresh-water plants: A potential source of protein. Econ. Bot. 4:359.

Boyd, C. E. 1973a. Summer algal communities and primary productivity in fish ponds. Hydrobiologia 41:357–390.

Boyd, C. E. 1973b. Amino acid composition of freshwater algae. Arch. Hydrobiol. 72:1–9.

Bryant, C. B. 1973. Control of aquatic weeds by mechanical harvesting. Pestic. Art. News Serv. 19:601–606.

Chavez, M. I., R. L. Shirley, and J. F. Easley. 1976. Effects of feeding hyacinths to rats for three generations. Soil Crop Sci. Soc. Fla., Proc. 35:74–76.

Chung, P., W. G. Pond, J. M. Kingsbury, E. F. Walker, Jr., and L. Krook. 1978. Production and nutritive value of *Arthrospira platensis*, a spiral blue-green alga grown on swine wastes. J. Anim. Sci. 47:319–330.

Clement, G., C. Giddey, and R. Menzi. 1967. Amino acid composition and nutritive value of the alga *Spirulina maxima*. Sci. Food Agric. 18:497–501.

Cook, B. B. 1962. The nutritive value of waste-grown algae. Am. J. Public Health. 52:243–251.

Culley, D. D., Jr., and E. A. Epps. 1973. Use of duckweed for waste treatment and animal feed. J. Water Pollut. Control Fed. 45:337–347.

Dam, R., S. Lee, P. C. Fry, and H. Fox. 1965. Utilization of algae as a protein source for humans. J. Nutr. 86:376–382.

Davis, I. F., M. J. Sharkey, and D. Williams. 1975. Utilization of sewage algae in association with paper in diets of sheep. Agric. Environ. 2:333–338.

Desai, M. C., and P. C. Shukla. 1975. Effect of feeding seaweed to lactating cows on body weights and milk production. Indian J. Anim. Sci. 45:823–827.

Golueke, C. G., and W. J. Oswald. 1965. Harvesting and processing sewage-grown planktonic algae. J. Water Pollut. Control Fed. 37:471.

Grau, C. R., and N. W. Klein. 1957. Sewage-grown algae as a feedstuff for chicks. Poult. Sci. 36:1046–1051.

Hart, M. R., D. de Fremery, C. K. Lyon, and G. O. Kohler. 1978. Dewatering kelp for fuel, feed, and food uses: Process description and material balances. Trans. ASAE 21:186–189.

Heffron, C. L., J. T. Reid, W. M. Hascheck, A. K. Furr, T. F. Parkinson, C. A. Bache, W. H. Gutenmann, L. E. St. John, Jr., and D. J. Lisk. 1977. Chemical composition and acceptability of aquatic plants in diets of sheep and pregnant goats. J. Anim. Sci. 45:1166–1172.

Herbert, J. G., and G. Wiener. 1978. The effect of breed and of dried seaweed meal in the diet on the levels of copper in liver, kidney and plasma of sheep fed on a high copper diet. Anim. Prod. 26:193–201.

Hillman, W. S., and D. D. Culley, Jr. 1978. The uses of duckweed. Am. Sci. 66:442–451.

Hintz, H. F., and H. Heitman, Jr. 1967. Sewage-grown algae as a protein supplement for swine. Anim. Prod. 9:135–140.

Hintz, J F., H. Heitman, Jr., W. C. Weir, D. T. Torell, and J. H. Meyer. 1966. Nutritive value of algae grown on sewage. J. Anim. Sci. 25:675–681.

Linn, J. G., E. J. Staba, R. D. Goodrich, J. C. Meiske, and D. E. Otterby. 1975a. Nutritive value of dried or ensiled aquatic plants. I. Chemical composition. J. Anim. Sci. 41:601–609.

Linn, J. G., R. D. Goodrich, D. E. Otterby, J. C. Meiske, and E. J. Staba. 1975b. Nutritive value of dried or ensiled aquatic plants. II. Digestibility by sheep. J. Anim. Sci. 41:610–615.

McDowell, R. E., and G. A. Leveille. 1963. Feeding experiments with algae. Fed. Proc. 22:1431–1438.

National Research Council. 1971. Atlas of Nutritional Data on United States and Canadian Feeds. Washington, D.C.: National Academy of Sciences.

National Research Council. 1976. Making Aquatic Weeds Useful: Some Perspectives for Developing Countries. Washington, D.C.: National Academy of Sciences.

Naylor, J. 1976. Production, Trade and Utilization of Seaweeds and Seaweed Products. FAO Technical Paper No. 159. Rome: Food and Agriculture Organization of the United Nations.

Pieterse, A. H. 1978. The water hyacinth (*Eichhornia crassipes*): A review. Abstr. Trop. Agric. (Roy. Trop. Inst., Amsterdam) 4(2):9–42.

Robinette, H. R., M. W. Brunson, and E. J. Day. 1980. Use of duckweed in diets of channel catfish. Thirty-Fourth Annual Conference, Southeast Association of Fish and Wildlife Agencies. Nashville, Tenn.: Association of Fish and Wildlife Agencies.

Rusoff, L. L., D. T. Gantt, D. M. Williams, and J. H. Gholson. 1977. Duckweed—A potential feedstuff for cattle. J. Dairy Sci. Suppl. 1. 60:161.

Rusoff, L. L., E. W. Blakeney, Jr., and D. D. Culley, Jr. 1980. Duckweed (Lemnaceae family): A potential source of protein and amino acids. J. Agric. Food Chem. 28:848.

Shukla, P. C., P. M. Talpada, and B. M. Patel. 1974. Utilization of seaweeds in the concentrate ration of growing calves. Indian J. Anim. Sci. 44:428–431.

Stephens, E. L., J. F. Easley, R. L. Shirley, and J. F. Hentges, Jr. 1973. Availability of nutrient mineral elements and potential toxicants in aquatic plant diets fed steers. Soil Crop Soc. Fla., Proc. 32:30–32.

Stephenson, W. A. 1973. Seaweed in Agriculture and Horticulture. 2nd ed. East Ardsley, England: E. P. Publishing.

Truax, R. E., D. D. Culley, M. Griffith, W. A. Johnson, and J. P. Wood. 1972. Duckweed for chick feed? La. Agric. 16:8–9.

Whittemore, C. T., and J. K. Percival. 1975. A seaweed residue unsuitable as a major source of energy as nitrogen for growing pigs. J. Sci. Food Agric. 26:215–217.

Appendix Tables

TABLE 1 Proximate Composition and Energy Value

	Composition of Dry Matter						Fiber					Ruminant			Swine	Poultry	
Type of Resource	Dry Matter (%)	Crude Protein (%)	DP (%)	NPN (%)	Ash (%)	Ether Extract (%)	Crude (%)	Lignin (%)	ADF (%)	NDF (%)	DDM (%)	TDN (%)	DE (Mcal/kg)	DE (Mcal/kg)	ME (Mcal/kg)	References
Activated sludge, fruit cannery		39.1			11.64	0.8	3.2									Esvett (1976)
Algae		50.9			10.5	6.6	4.9				54.0	62.2	3.14			NRC (1971)
		31.0–44.4														Grau and Klein (1957)
		41.8				7.2										Cook (1962)
		53.0				9.2										McDowell and Leveille (1963)
		58.2				2.9	6.0									Dam et al. (1965)
		50.9				6.0	6.2						2.6			Hintz et al. (1966)
		49.1									54.2					
		68.4									59.5					
Almond hulls—range (max., min.)	70–90	4.3			5.7	7.3	0.6					72				Anonymous (1970)
	≥87				4–7	1–4	10–17									Hamilton (1975)
	90				≤9		≤15									Hamilton (1975)
	88.4	4.3			6.6	4	10.8					74.9				Hamilton (1975)
	91.0	4.4	0.43		6.7	4.1	11.0		28			57	2.51			Morrison (1956)
							15.0									NRC (1978)
Apple pomace (wet)	28.6	5.9			2.0	7.4	27.6					75.8				Fontenot et al. (1977)
	21.1	6.2	2.4		4.3	6.2	17.5		29.6			72				Morrison (1956)
Without rice hulls		4.9	1.8				17.0		53.8			44.0				Adams (1976)
With rice hulls		4.7	1.7				31.0					67.7				Adams (1976)
Apple pomace (dry)	89.4	5.0	1.9		2.5	5.6	17.4					69.0				Morrison (1956)
	89	4.9	1.8				17.0					72.0				NRC (1978)
	90	7.3			2.6	4.6	22.6									Bath et al. (1978)
Apple pomace (dehydrated)	89.0	4.9					17					69	3.04			NRC (1978)
Apple pectin pulp (wet)	16.7	9.0	3.6		3.6	5.4	34.7					58.1				Morrison (1956)

Feed									Reference
Apple pectin pulp (dry)	91.2	7.7		3.6	8.0	26.5		60.3	Morrison (1956)
Aquatic plants		17.1	2.9		2.1	20.6	40.1	43.6	Linn et al. (1975b)
		13.0			1.5	19.5	32.4		Linn et al. (1975a)
Asparagus butts		17.2			1.1	35.1		51.7	Folger (1940)
	91.0	17.1		7.7	1.1	35.0		52.2	Bath et al. (1978)
Aspen, bark, ensiled	60	2.2		2.7	5.8	53.7	59.1	37	Enzmann et al. (1969)
Aspen, steamed, stake process	55–60	1.0		1.0		54.0		55–60 55	Stake Tech. Ltd. (1979) 1.6
									1.3 NE_m
									0.88 NE_g
									1.3 NE_l
Aspen, whole tree	40.8	1.2		3.2			64.9		Singh (1978)
Aspergilis fumagatus		40				25.4			
Bagasse							10.8 49.2	32.0	Rexen (1977)
Barley straw	87.0	3.8		4.3			5.5 49.0 83.4	46.5	Wilkinson and
							83.6		Santillana (1978)
	91.6	3.6		4.4			8.2 57.5 87.3	55	Coombe et al. (1979b)
							7.9 51.3 79.9	45.3	Rexen (1977)
	86.2	4.4		6.1		46.8			Anderson (1978)
	88.5	3.87			1.6	35.3		57.6	Horton and Steacy (1979)
Beet sugar tops	12.9	14.8		5.4	0.3	11.9			Leonard et al. (1973)
Birch, foliage (muka)		8.0		4.2	8.8			58.4	Keays (1976)
Broccoli	9.4	28.7		9.6	3.2	10.6			T. Canavan, General Foods Corp., personal communication, 1979
Broccoli leaf meal	94.25	39.5				8.5			Ben-Gera and Kramer (1969)
Broccoli stems	47.1	23.2				16.7			Ben-Gera and Kramer (1969)
Broiler litter	84.3	38.9		11.7	5.6	17.5			Harmon et al. (1974)
	84.1	24.8	1.63					67.6	Caswell et al. (1975)
	87.3	25.8	1.89					67.1	Caswell et al. (1975)
	82.9	26.8		16.7	1.5	23			Harmon et al. (1975a)
	81.4	31.5	2.59	19.4	3.5	19.5			Caswell et al. (1978)
	77.7	40.0	3.65	13.8	2.4	19.2			Caswell et al. (1978)
		20.1	1.04	22	3.4				Cross et al. (1978)

229

TABLE 1 (*Continued*)

Type of Resource	Dry Matter (%)	Crude Protein (%)	DP (%)	NPN (%)	Ash (%)	Ether Extract (%)	Crude Fiber (%)	Lignin (%)	ADF (%)	NDF (%)	DDM (%)	TDN (%)	Ruminant DE (Mcal/kg)	Swine DE (Mcal/kg)	Poultry ME (Mcal/kg)	References
Broiler litter (processed)	85	22.5			22.3	1.0	27.6									Cullison et al. (1976)
	85	24.9			32.5	0.7	19.4									Cullison et al. (1976)
	90.3	39.1		3.98	14.7	3.3	16.5								1.274	Blair and Herron (1982)
	89.1	32.0		2.79	17.9	2.8	15.1					60	2.212			Bhattacharya and Fontenot (1966)
	88.9	30.6		2.72	19.0	2.8	14.6					59.5	2.15			Bhattacharya and Fontenot (1966)
	78.7	32.2		1.66	13.4	3.1	19.8									Perez-Aleman et al. (1971)
	87.4	34.7			11.6	4.5	16.8									Harmon et al. (1974)
	99.7	34.1			11.2	5.8	18.5									Harmon et al. (1974)
	99.1	37.2			12.2	5.5	17.2									Harmon et al. (1974)
	85.3	25.3		1.66												Caswell et al. (1975)
	84.4	26.3		2.0												Caswell et al. (1975)
Broiler waste (processed)	92.6	28.2									66.5				1.84	Yoshida and Hoshii (1968)
	80.3	35.9				3.62										Kubena et al. (1973)
	78.3	40.8				4.0										Kubena et al. (1973)
	77.0	40.8				3.6										Kubena et al. (1973)
	92.9	47.8		5.04	10.8	3.2	10.8									Bhargava and O'Neil (1975)
	93.1	34.6		3.37	17.7	1.8	17.1									Bhargava and O'Neil (1975)
		25.3		2.19												Trakulchang and Balloun (1975a)

Brussels sprouts	18.0	43.1		24.0				Smith and Calvert (1976)
		9.4		10.6	1.7	21.7		T. Canavan, General Foods Corp., personal communication, 1979
Cacao shell		13.5						Chatt (1953)
	95.1	16.0	4.4	6.5	3.0	16–20		Morrison (1956)
Cage layer waste	31.8	31.9		10.8	3.2	17.4		Smith et al. (1980)
	24.0	20.1		25.5	2.3			Silva et al. (1976)
		23.4	1.97	17.8	1.6	21.7	84.5	Van Soest and Robertson (1976)
	30.1	43.1	5.44	24.0	2.0	16.5		Evans et al. (1978a)
	31.5	33.1	2.86	18.1	1.7			Evans et al. (1978a)
	35.0	48.8		30.3				Evans et al. (1978b)
	33.0	43.0		17.2				Evans et al. (1978b)
	29.5	26.2	3.49	19.7	2.8		72.1	Smith et al. (1978)
	30.5	35.7	4.48	26.6	2.3		79.9	Smith et al. (1978)
Cage layer waste (processed)	94.0	42.3	4.68	29.0	2.3			Swingle et al. (1977)
	86.2	31.9	2.75	24.8	1.2			Blair and Herron (1982) 1.218
	89.5	31.3	3.13	26.8	3.2			Blair and Herron (1982) 1.063
	9.14	23.5		33.0		16.8		Hamblin (1980)
	89.5	26.8		39.1	1.61	14.0		Helmer (1980)
	85.5	32.4		36.5	3.04	14.3		Helmer (1980)
	93.9	39.8		31.8	3.24	12.9		Helmer (1980)
	89.1	26.5		41.6	1.12	9.7		Helmer (1980)
		18.6	1.18		1.44	14.0		Chang et al. (1974)
		17.2	1.09		0.96	19.7		Chang et al. (1974)
	96.1	35.7	3.85	21.2	2.0	19.8		McNab et al. (1974) 0.836
		35.6	3.41	25.7	2.02	11.1		Chang et al. (1975)
	95.0	29.3	2.5		3.2	14.7		Stapleton and Biely (1975)
		26.8	1.9		3.3	12.6		Stapleton and Biely (1975) 1.158
		27.2	2.0		3.2	11.9		Stapleton and Biely (1975) 1.158

231

TABLE 1 (Continued)

Type of Resource	Dry Matter (%)	Crude Protein (%)	DP (%)	NPN (%)	Ash (%)	Ether Extract (%)	Crude Fiber (%)	Lignin (%)	ADF (%)	NDF (%)	DDM (%)	TDN (%)	DE Ruminant (Mcal/kg)	DE Swine (Mcal/kg)	ME Poultry (Mcal/kg)	References
Cage layer waste (processed)	87.0	37.6		4.4	34.1	1.8									1.03	Shannon et al. (1973)
	86.4	31.9		3.42	32.5	2.0									0.99	Shannon et al. (1973)
	88.0	21.9		1.55	35.8	3.4									1.26	Shannon et al. (1973)
	92.8	38.5		3.94	23.2	2.0									1.27	Shannon et al. (1973)
	84.4	35.0		3.23	32.6	1.1									1.04	Shannon et al. (1973)
	96.1	38.3		4.26	20.7	2.4									0.75	Shannon et al. (1973)
		29.4		2.71		1.59	10.97									Chang et al. (1974)
		31.2		2.88		1.76	10.96									Chang et al. (1974)
		30.45		3.12		1.72	12.29									Chang et al. (1974)
		20.76		1.56		4.04	16.52									Chang et al. (1974)
															1.1	Pryor and Connor (1964)
	80.8	28.8			15.9	1.8	7.5									Parigi-Bini (1969)
	92.3	28.7			16.7	1.73	11.6									Lowman and Knight (1970)
	92.2	28.8			18.1	1.52	8.9				56.6				1.88	Lowman and Knight (1970)
	81.5	36.9			27.4	1.43										Bull and Reid (1971)
	85.0	32.6			27.6	1.75									1.912	Hodgetts (1977)
	91.2	19.7		1.26	29.2	4.1	11.9									Polidori et al. (1972)
	82.6	18.4		1.18	49.4	1.5	14.2								1.08	Shannon et al. (1973)
	82.3	31.6		3.26	40.1	1.7									0.64	Shannon et al. (1973)
Carrots, fresh	11.9	9.6			8.8	3.5	11.4									Leonard et al. (1973)
Carrots (wet)	10.1	13.9			7.9	<1	9.9									T. Canavan, General Foods Corp., per-

Cassava leaves		14.8	11.1	7.0		15.2	sonal communication, 1979
							Ross and Enriquez (1969)
Cattle waste (processed)	24.8	18.9	11.5	1.1		13.4	Anthony (1971)
	25.6	20.3	36.4		68.3		Anthony (1971)
		10.6			50.7		Lipstein and Bornstein (1971)
		12.4	26.1	1.6		23.5	Lipstein and Bornstein (1971)
	10	18.4	5.4	2.8	44.8	43.3	31.4	Blair and Knight (1973)
		13.2	14.0	2.6		16.6	30.5	Lucas et al. (1975) (Ruminant ME 0.485)
		14.8			45.8	33.1		Thoriacius (1976)
		13.2			55.7	22.0		Van Soest and Robertson (1976)
					35.6	34.1		Van Soest and Robertson (1976)
		25.0			26.1	66.2		Van Soest and Robertson (1976)
		16.9			33.6	47.8		Van Soest and Robertson (1976)
	15.7	21.5	1.6	7.4			22.7	Schake et al. (1977)
	22.7	17.2	8.1	3.3			23.3	Lamm et al. (1979)
	23.6	17.2	9.1	3.2			22.6	Lamm et al. (1979)
Cauliflower	7.6	22.4	14.5	1.3			13.1	T. Canavan, General Foods Corp., personal communication, 1979
Cauliflower leaves		32.5	0.4	5.9			8.8	Livingston et al. (1972)
Celery	7.2	23.6	13.9	1.4			8.3	T. Canavan, General Foods Corp., personal communication, 1979

TABLE 1 (*Continued*)

Type of Resource	Dry Matter (%)	Crude Protein (%)	DP (%)	NPN (%)	Ash (%)	Ether Extract (%)	Crude Fiber (%)	Lignin (%)	ADF (%)	NDF (%)	DDM (%)	TDN (%)	Ruminant DE (Mcal/kg)	Swine DE (Mcal/kg)	Poultry ME (Mcal/kg)	References
Celery tops		25.3			11.6	3.0	15.3					79.5				Haines et al. (1959)
Citrus activated sludge	97.3	41.2					13.4									Damron et al. (1974)
Citrus seed meal																
Commercial	85	39.9			7.0	6.7	8.8									Driggers et al. (1951)
Detoxified	87.7	49			9.2	0.1	10.9									Driggers et al. (1951)
Citrus meal		7.3			6.7	3.3	14.0									Hendrickson and Kesterson (1965)
Citrus molasses	71	5.8			6.6	0.28										Hendrickson and Kesterson (1965)
Citrus pulp	90.1	6.5			7.7	3.4	12.76									Hendrickson and Kesterson (1965)
Corn																
Cobs		6.8						5.5	40.2	71.5	62.3					Lamm (1976)
Stalks		6.6						6.3	49.9	61.5	62.1					Lamm (1976)
Husks and leaves		7.3						7.2	41.7	68.4	66.2					Lamm (1976)
Cobs		5.6						23.6	52.5	85.8	37.7					Lamm (1976)
Stalks		5.8						7.5	50.9	73.3	48.8					Lamm (1976)
Husks and leaves		7.6						21.5	48.7	78.6	47.9					Lamm (1976)
Corncob	58	2.8			1.4			7.4	42.8	80.1	60.0					Vetter (1973)
Corn husk	55	2.8			3.4			6.7	41.6	78.0	68.0					Vetter (1973)
Corn husklage	78	3.7			3.3						65.0					Vetter (1973)
Corn leaf	76	7.0			13.7			5.1	37.0	61.8	58.0					Vetter (1973)
Cornstalk	31	3.7			4.7			10.5	47.1	70.8	51.0					Vetter (1973)
Corn stalklage, late harvest									51.9	77.1	47.2					Berger et al. (1979)

Corn stalklage, early harvest								
Corn (sweet) waste	19.7	6.6	3.5	1.0		24.4		45.2	70.1	52.1	Berger et al. (1979); T. Canavan, General Foods Corp., personal communication, 1979
Cotton gin trash	92.5	5.6			2.9		13.8	55.1			Brown et al. (1979)
	91.8	10.2	7.7		3.5	39.7					Holloway et al. (1974)
	95.7	4.8			2.8	69.6					Brown et al. (1977)
Cottonseed hulls		5.4	3.2		1.6	47.0	21.8	64.5			Hale et al. (1969)
		3.9			1.0	41.5					Heinemann (1976)
Date seeds		6.6	3.2		9.0	15.7				74.6	El Shazly et al. (1963)
Dried whey, range	2–6	12–14	8–12		.3–5						Jones (1974)
Duckweed		31.7			5.2	8.2					Truax et al. (1972)
		17.4			2.7	10.4				1.96	Culley and Epps (1973)
Faba bean residue		9.6	12.6					44.1		61.4	Kernan et al. (1980)
Fermented, ammoniated whey	55–65	45	4.7								Juengst (1979)
Field pea residue		7.9	6.4					50.6		50.4	Kernan et al. (1980)
Fir, foliage (muka)		6.6		10.4				33.2			Gerry and Young (1977)
Grain sorghum											
Leaf	66	10.0	10.6				7.3	39.6	64.8	56	Vetter (1973)
Stalk	25	3.6	8.7				10.1	17.2	64.8	57	Vetter (1973)
Silage	32	6.8	10.4						70.0	53	Vetter (1973)
Stover	40	4.7	10.2				9.3	50.0		54	Vetter (1973)
Stalk		6.6					4.9	41	78	52	Ward et al. (1979)
Stalk, weathered		5.1					6.3	50	89	47	Ward et al. (1979)
Stalk, grazed		5.6					6.2	49	89	47	Ward et al. (1979)
Leaf		12.4					5.4	37	75	49	Ward et al. (1979)
Leaf, weathered		9.5					7.2	44	80	40	Ward et al. (1979)
Leaf, grazed		8.7					7.8	46	83	37	Ward et al. (1979)
Grape pomace, average	90.6	13.4	6.6		8.29	30.96					Folger (1940)
	90.6	14			7.9	33.3					Prokop (1979)
		11–14	5–10		5–9	17–40					Amerine et al. (1972)

TABLE 1 (Continued)

Type of Resource	Dry Matter (%)	Crude Protein (%)	DP (%)	NPN (%)	Ash (%)	Ether Extract (%)	Crude Fiber (%)	Lignin (%)	ADF (%)	NDF (%)	DDM (%)	TDN (%)	Ruminant DE (Mcal/kg)	Swine DE (Mcal/kg)	Poultry ME (Mcal/kg)	References
Grass straws																
Bluegrass	92.4	6.0			5.8			5.8	47.4							Durham and Hinman (1979)
	93.7	5.5			6.0	1.1	38.4	9.8	50.2	78.3	33.7					Early and Anderson (1978)
Fescue		8.9			4.4	1.9	37.5	13.0	43.7		31.0					Guggolz et al. (1971)
Perrennial rye		5.1			5.4	1.6	44.2	13.7	53.0		29.4					Guggolz et al. (1971)
Bentgrass		5.5			6.7	1.8	41.2	13.4	50.6		39.6					Guggolz et al. (1971)
		4.6			2.7	1.9	39.4	11.3	45.0		34.0					Guggolz et al. (1971)
Annual rye		4.8			6.3	1.4	42.0	2.7	49.7		31.9					Guggolz et al. (1971)
Hardwood, hydrolyzed (Jelks process)	60	1.7			1.6								0.75			J. W. Jelks, Sandy Springs, Okla., personal communication, 1979
Lettuce, fresh	5.3	22.6			17.0	5.7	13.2									Leonard et al. (1973)
Oat straw	90.8	3.2			7.5	1.9	41.1				54.5					Anderson (1978)
											58.5					Rexen (1977)
	87.4	2.6					42.4									Horton and Steacy (1979)
Paunch manure (dried)	84.7	12.2			7.9	5.2	25									Ricci (1977)
Pea, aerial part without seeds		11.7			9.0	2.7	26.0									NRC (1971)
Peanut hulls	90.5	8.4			3.6	1.8	63.5	29.9	68.3	77.2		63.4				Utley and McCormick (1972)
Pear molasses	76.4	1.6			6.9							86				Guilbert and Weir (1951)

Pear pulp	92.4	5.8		5.8	3.1	34.1		Guilbert and Weir (1951)
Pecan shells		3.13		2.0	4.3	50.7		Cullison et al. (1973)
Penicillium chrysogenun		31.9		19.8	7.0			T. Canavan, General Foods Corp., personal communication, 1979
Peppers	6.6	21.2		13.6	3.0	21.2		
Pimento waste (dried)		11.9			9.48	44.5		Livingston et al. (1974)
Ponderosa pine, bark		1.9		3.8			64.5	Kamstra et al. (1980)
Ponderosa pine, sawdust		1.2		0.2			79.5	Kamstra et al. (1980)
Potato starch waste								
Wet pulp	16.63	7.2			0.36	9.62		Brugman and Dickey (1955)
Dry pulp	87.7	8.8			0.44	7.0		Brugman and Dickey (1955)
Whole potato	19.0	9.9			0.32	3.05		Brugman and Dickey (1955)
Potato waste, symba yeast	94	51.1		5.8	3.2	1.1		Skogman (1976)
Prunes		6.7			2.7	14.1	77.5	Folger (1940)
Pulp fines, sulfite mill hardwood	23.9	1.2		1.5	3.0		86.4	Lemieux and Wilson (1979)
Raisins, dried	84.8	4.0		3.5	1.1	5.2	48.9	Leonard et al. (1973)
Raisin pulp	88.68	10.8			11.9	21.8	57.5	Mead and Guilbert (1926)
Rice hulls		3.8		22.2			17.2 77.9 85.0	McManus and Choung (1976)
Rice straw							6.3 45.9 80.7 46.3	Rexen (1977)
Rape straw							9.6 54.8 75.3 34.5	Rexen (1977)
Rye straw		4.1		5.2			50.2 40.0	Kernan et al. (1980)
Seaweed		10.9		24.2	1.0	9.2	43.2 1.91	Whittemore and Percival (1975) 22MJ/kg

237

TABLE 1 (Continued)

Composition of Dry Matter

Type of Resource	Dry Matter (%)	Crude Protein (%)	DP (%)	NPN (%)	Ash (%)	Ether Extract (%)	Crude Fiber (%)	Lignin (%)	ADF (%)	NDF (%)	Ruminant DDM (%)	Ruminant TDN (%)	Swine DE (Mcal/kg)	Swine DE (Mcal/kg)	Poultry ME (Mcal/kg)	References
Soybean pod	88	6.1			7.3			8.7	40.6	55.7	51					Vetter (1973)
Soybean stalk	88	4.0			2.9			18.4	63.8	76.5	35					Vetter (1973)
Soybean stover	87	4.3			4.8			15.6	61.5	75.3	40					Vetter (1973)
Soybean straw		5.6						19.1	56.8	77.9	42.9					Gupta et al. (1978)
Spruce, foliage (muka)		8.8			4.4	6.5										Keays (1976)
		6.4			8.1											Gerry and Young (1977)
St. Louis SCW-I	81.3	5.9			21.5				36.0							Belyea et al. (1979)
St. Louis SCW-II	83.2	7.0			23.7				72.2							Belyea et al. (1979)
Streptomyces rimosus							6.3		70.5							
Streptomyces viridifaciens		24.4			47.2	15.6	2.9	20.1								Martin et al. (1976)
					3.4											Butterworth (1962)
Sugarcane bagasse	25.6	6.3			6.2	2.2	35.0					59.3				Pate et al. (1971)
Sugarcane tops	31.7	6.0			7.4	1.7	31.2									Pate and Coleman (1975)
	39.7	5.8			6.2	1.5	33.9									
Sunflower residue		6.7			12.1				26.9		70.5					Kernan et al. (1980)
Sweet corn Aerial part, no ears		7.3			6.2	1.8	26.0					65.5				NRC (1971)
Whole plant, mature		11.7			7.2	3.0	23.2					66.7				NRC (1971)
Cannery residue		8.9			3.3	2.3	22.5					70.1				NRC (1971)
Sweet potatoes	31.8	5.03	0.63		3.8	1.3	6.0					80.5				Morrison (1956)
Sweet potatoes, dried		2.8	0.44		6.7	0.33	10.6					76.5				Morrison (1956)
Sweet potato meal	90.2	5.4	0.78		4.5	1.0	3.7					80.6				Morrison (1956)

Sweet potato trimmings, dried		5.5		13		16.6		Bond and Putnam (1967)
Sweet potato vine meal		13.2		15.8	2.8			Garlich et al. (1974)
Swine waste (processed)		27.7						Harmon et al. (1972)
		49.0						Harmon et al. (1973)
		18.9		18.1	6.6		26.5	Pearce (1975)
		23.5	1.16	15.3	8.0	14.8		Kornegay et al. (1977)
		20.5	0.55	20.5	6.45	15.2		Kornegay et al. (1977)
Tannery waste								
Waste hair	38	61.6						Komanowsky and Craig, personal communication, 1979
Fleshings	20	50						Komanowsky and Craig, personal communication, 1979
Trimmings	25	84						Komanowsky and Craig, personal communication, 1979
Splits	25	90						Komanowsky and Craig, personal communication, 1979
Tomato pomace, dried	94.7	23.9	16.9	3.5	15.3	32.2		Bath et al. (1978)
Tomato pomace, dehydrated	92	23.9					50	NRC (1978)
Tomato pulp	89.1	24.0		10.6	4.2	26	58 2.56	Ammerman et al. (1963)
Triticale straw	94.2	2.7		6.4		17.8		
Washington Post		18		0.4			58.3 37.5	Keman et al. (1980)
Water hyacinth	3.46	15.8		18		94	81	Mertens et al. (1971)
	5.9	16			3.5			Baldwin et al. (1974)
	3.8	11.8					43.2 39.6	Bagnall et al. (1974)
								Baldwin et al. (1975)

TABLE 1 (*Continued*)

Type of Resource	Dry Matter (%)	Crude Protein (%)	DP (%)	NPN (%)	Ash (%)	Ether Extract (%)	Crude Fiber (%)	Lignin (%)	ADF (%)	NDF (%)	DDM (%)	TDN (%)	Ruminant DE (Mcal/kg)	Swine DE (Mcal/kg)	Poultry ME (Mcal/kg)	References
Wheat straw	92.5	4.1			10.2	1.7	41.1	6.9	54.0	86.0	47.7					Anderson (1978)
								7.1	49.1	80.5	45.1					Coombe et al. (1979a)
																Rexen (1977)
	88.0	2.52					43.7				51.2					Horton and Steacy (1979)
Whey dried whole	93	13.1	7		9.0											Schingoethe (1976)
Whey solids coagulated with chitosan	93.2	77.6			10.2	0.16										Bough and Landes (1976)

TABLE 2 Mineral Composition

Type of Resource	Ash (%)	Ca (%)	P (%)	Na (%)	Cl (%)	Mg (%)	K (%)	S (%)	Fe (mg/kg)	Cu (mg/kg)	Co (mg/kg)	Mn (mg/kg)	I (mg/kg)	Zn (mg/kg)	Se (mg/kg)	Mo (mg/kg)	Fl (mg/kg)	References
Algae	10.5	1.39	2.12	0.23		1.60	0.92		2300	100		300		1800				Cook (1962)
	19.1	2.94	2.84						2600									McDowell and Leveille (1963)
	4.8																	Dam et al. (1965)
	7.2	0.22	1.77															Hintz et al. (1966)
	6.2	1.93	2.22	0.23		1.60	0.92		2300	100		300		1800				Davis et al. (1975)
										260								Anonymous (1970)
Aquatic plants	5.7	0.08	0.88	0.70	0.23	0.31	1.61		560									Linn et al. (1965a)
	2.1	1.62	0.27	0.32		0.28	1.51		965			269				20		Linn et al. (1965b)
	26.1	3.36	0.32						582							315		
		5.53		0.73	0.75		2.15		991	8.1	1.3	170	5.3	43	0.3	3	17.9	Heffron et al. (1977)
Aspen, bark	4.3	1.09	0.035	0.16		0.06	0.22	<0.01	194	21		36		68				A. J. Baker, Forest Products Laboratory, Madison, Wis., personal communication, 1978
Aspen, bark ensiled		1.2	0.03	0.003		0.09	0.22		74.1	7.7		21.0		140		0.1		Enzmann et al. (1969)
Aspen, sulfite parenchyma cell fines (unbleached)	2.1	0.6	0.02	0.02		0.08	0.01	0.37	340	40		7		37				Baker et al. (1975)
Aspen, sulfite screen rejects (mill 2)	2.1	0.7	<0.01	0.02		0.04	0.05	1.1	63	6		2		6	<0.04			Baker et al. (1975)
Aspen, whole tree								<0.01		4.9					<0.04			A. J. Baker, Forest Products Laboratory, Madison, Wis., personal communication, 1978
Aspen, wood	0.8	0.18	0.003	0.008		0.03	0.06	<0.01	35	6		10		19				Baker et al. (1975)

TABLE 2 (Continued)

Type of Resource	Ash (%)	Ca (%)	P (%)	Na (%)	Cl (%)	Mg (%)	K (%)	S (%)	Fe (mg/kg)	Cu (mg/kg)	Co (mg/kg)	Mn (mg/kg)	I (mg/kg)	Zn (mg/kg)	Se (mg/kg)	Mo (mg/kg)	Fl (mg/kg)	References
Aspergilis fumagatus		0.01	0.82															Anderson (1978)
Barley straw		0.72	0.05															
Barley straw and chaff		0.21	0.13			0.14	1.52		88	4.1		25		13.7				Anderson (1978)
Birch, foliage (muka)		0.8	0.3			0.3	0.7		101	8		30		121				Keays (1976)
Broiler litter	11.7	1.5	1.26															Harmon et al. (1974)
	22	3.1	2.1															Cross et al. (1978)
Broiler litter (processed)	14.7	2.1	1.44	0.33			2.55											Blair and Herron (1982)
	17.9	2.8	2.9															Bhattacharya and Fontenot (1966)
	19.0	2.5	2.3															Bhattacharya and Fontenot (1966)
	13.4	3.05	1.96			0.54												Cenni et al. (1969)
	13.4	2.76		0.42		0.38				23				343				Perez-Aleman et al. (1971)
	11.6	1.5	1.26															Harmon et al. (1974)
	11.2	1.4	1.26															Harmon et al. (1974)
	12.2	1.4	1.26															Harmon et al. (1974)
	22	3.1	2.1															Cross et al. (1978)
Broiler waste (processed)	10.8	1.44	1.51															Bhargava and O'Neil (1975)
	17.7	3.63	2.35															Bhargava and O'Neil (1975)
		5.12	1.97	0.47		0.53	1.38		1690	32		432		326				Trakulchang and Balloun (1975a)
	24	3.4	2.2															Smith and Calvert (1976)

Cage layer waste	17.8	3.8		0.28	0.5	1.91	1216	31	232	276		Silva et al. (1976)
	24.0	7.17	1.72	0.24	0.57	1.74	613	20	199	242		Evans et al. (1978a)
	18.1	5.1	1.62									Evans et al. (1978a)
	26.6	10.9	1.7	0.34	0.5	1.8	686.8	33.3	294.5			Smith et al. (1978)
	19.7	8.3	1.3	0.31	0.5	1.6	437.1	33.5	217.5			Smith et al. (1978)
	25.5	7.2	2.5	0.21	0.49	3.2	357.8	37.7	112.8	108.8		Smith et al. (1980)
Cage layer waste (processed)	31.0	7.0	1.91					35			0.27	Helmer (1980)
	37.8	9.67	2.41					45.5			0.85	Helmer (1980)
			2.0		0.44	1.88	1187	66.2	457.8	717.2		Stapleton and Biely (1975)
	24.8	5.3	2.0	0.39		1.87						Blair and Herron (1982)
	26.8	7.8	2.1	0.45		1.45						Blair and Herron (1982)
	33.0	8.0	2.4	0.42	0.63	2.3	2000	51	291	325	1.17	Hamblin (1980)
	39.1	9.39	2.67					50.3			0.06	Helmer (1980)
	36.5	9.59	2.32					42.1			0.56	Helmer (1980)
	31.3	8.31	2.34					36.2			0.62	Helmer (1980)
	41.6	12.68	2.58					64			0.36	Helmer (1980)
	35.8	9.5	3.2	0.22	1.16	1.33	1400	55	318	318		Shannon et al. (1973)
	23.2	5.5	1.9	0.11	0.74	1.53	1400	51	232	226		Shannon et al. (1973)
	32.6	7.0	2.1	0.48	0.71	1.68	2100	94	409	320		Shannon et al. (1973)
	20.7	5.1	1.9	1.19					271	352		Shannon et al. (1973)
		6.72	3.0	0.88								Chang et al. (1974)
		9.48	2.2									Chang et al. (1974)
		10.17	2.3									Chang et al. (1974)
		7.05	2.47									Chang et al. (1974)
		6.42	2.67									Chang et al. (1974)
		7.41	3.14									Chang et al. (1974)
	25.7	8.5	2.2									Chang et al. (1975)
			1.5	0.6	0.9	0.52	874	70.5	842	895		Stapleton and Biely (1975)
		2.3		0.68	1.0	0.52	1010	45.3	632	978		Stapleton and Biely (1975)

TABLE 2 (Continued)

Type of Resource	Ash (%)	Ca (%)	P (%)	Na (%)	Cl (%)	Mg (%)	K (%)	S (%)	Fe (mg/kg)	Cu (mg/kg)	Co (mg/kg)	Mn (mg/kg)	I (mg/kg)	Zn (mg/kg)	Se (mg/kg)	Mo (mg/kg)	Fl (mg/kg)	References
	15.9	5.73	2.39															Parigi-Bini (1969)
	16.7	2.69	0.86	0.54	0.86	0.22			975	56	1.08	280		571				Lowman and Knight (1970)
	17.1	5.2	1.57	0.63	0.97													Bull and Reid (1971)
	27.4	9.3	1.5						683	31.4	1.08	265		467				Hodgetts (1971)
	27.6	7.88	2.39	0.53	0.83												30	Polidori et al. (1972)
	29.2	3.6	1.6	0.55			2.04		2200	300	3.5	230		400			40	Stapleton and Biely (1975)
	49.4	15.1	3.0	0.53	1.46	0.91	2.14		5000	67		490		508				Shannon et al. (1973)
	40.1	12.8	2.1	0.77	1.69	0.76	1.29		1200	74		344		389				Shannon et al. (1973)
	34.1	11.3	2.1	0.22	0.64	0.57	0.96		1500	63		218		287				Shannon et al. (1973)
	32.5	9.6	2.2	0.27	0.99	0.61	1.14		3300	109		325		405				Shannon et al. (1973)
Cattle waste																		
(processed)	11.5	0.87	1.6			0.4	0.5		1340	631		147.48		242.42				Anthony (1971)
	26.1	2.3		0.88	1.32													Lipstein and Bornstein (1971)
	36.4	4.9	0.7															Lipstein and Bornstein (1971)
	1.6	1.3	0.78															Schake et al. (1977)
Corn																		
Stalk		0.28	0.11	0.11		0.23	1.91	0.02	170	10		19		27				Vetter et al. (1973)
Leaf		0.80	0.18	0.11		0.28	1.23	0.32	400	16		60		32				Vetter et al. (1973)
Husk		0.19	0.08	0.11		0.13	1.09	0.12	150	12		20		19				Vetter et al. (1973)
Cob		0.04	0.06	0.11		0.08	0.85	0.06	200	12		6		24				Vetter et al. (1973)
Husklage		0.16	0.08	0.08		0.22	0.90		100	14		16		24				Vetter et al. (1973)
Stalklage		0.37		0.12		0.24	0.92	0.12	800	12		52		24				Vetter et al. (1973)
	11.4	0.72	1.71															Truax et al. (1972)
Duckweed	16.0	1.15	0.61			0.30	2.47		8000	13.8		9600		218				Culley and Epps (1973)

Hardwood, mixed, screen rejects (mill 1)	11.8	0.83	1.89		0.25	2.08	4700	13.3	1500	119	Culley and Epps (1973)
Hardwood, mixed, sulfite pulp fines (unbleached)	17.4	0.26	0.04	0.03	0.04	0.3	1.5	74	44	14	Baker et al. (1975)
Milo	1.8	0.21	<0.01	0.43	<0.01	<0.02	0.28	2300	8	9	4	Baker et al. (1975)
Stalk		0.29	0.08	0.08	0.26	0.17		95	14	48	22	Vetter et al. (1973)
Leaf		0.64	0.19	0.08	0.30	0.42		130	11	59	42	
Stover		0.48	0.24		0.26	0.32		440			46	
Penicillium cyso-genum	19.8	13.2	0.7	2.5	0.02	3.6						
Pine, southern, unbleached Kraft pulp fines (mill 5)	3.4	0.28	0.23	0.20	0.08	0.10	0.62	350	99	330	330	Baker et al. (1975)
Pulp fines, sulfite mill, hardwood		0.08	0.02	0.21	0.01	0.02	0.27	30	7	9	9	Lemieux and Wilson (1979)
Seaweed	24.2	1.50	0.27		0.71	4.40		10	9	7.3	0.033	NRC (1971)
Soybean												
Stalk		0.73	0.09		0.30	0.12						
Pod		1.02	0.17		0.58	0.04						
Stover		0.90	0.10		0.36	0.08		600	10	40		Vetter et al. (1973)
Spruce, foliage (muka)												
St. Louis SCW-I	21.5	>1		0.00	>1			>1%				Belyea et al. (1979)
St. Louis SCW-II	23.7	>1		1	>1			>1%				Belyea et al. (1979)
Streptomyces rimosus		1.2	0.09	0.03	0.012	0.015		42	6	30	7.5	
Streptomyces viridifaciens	47.2	0.19	0.16	12	0.15	8.8		37	7.5	60	14	
Swine waste (processed)		1.6	1.5									Harmon et al. (1972)
		3.33	3.83	2.75	1.49	4.14		5507	71	342	1148	Harmon et al. (1973)
	18.1	3.2	2.5		0.8	1.2	0.3	1940	249		526	Pearce (1975)
									6.1		0.3	

TABLE 2 (Continued)

Type of Resource	Ash (%)	Ca (%)	P (%)	Na (%)	Cl (%)	Mg (%)	K (%)	S (%)	Fe (mg/kg)	Cu (mg/kg)	Co (mg/kg)	Mn (mg/kg)	I (mg/kg)	Zn (mg/kg)	Se (mg/kg)	Mo (mg/kg)	Fl (mg/kg)	References
	15.3	2.7	2.1			0.93	1.34			62.8				530.4				Kornegay et al. (1977)
	20.5	4.39	3.04			1.17	0.91			73.9				633				Kornegay et al. (1977)
Washington Post	0.4	0.05	<0.05	0.1		0.02	<0.12		165	10		80		215				Mertens et al. (1971)
Water hyacinth	22.4	1.71	0.88			0.68	8.71											Baldwin et al. (1974)
	17.0	1.00	0.42			1.10	4.40											Bagnall et al. (1974)
Wheat chaff	0.05	0.19				0.12	1.00		209.4	3.33		54.9		5.6				Anderson (1978)
Wheat straw		0.17	0.04			0.10	1.61		126.7	3.2		37.2		3.2				Anderson (1978)

TABLE 3 Additional Mineral Elements (mg/kg)

Type of Resource	Al	Cd	Cr	Au	Hf	La	Pb	Hg	Ni	Sc	Ta	Sn	References
Algae	1200	10					100	1.0					Hintz et al. (1966)
							160						Davis et al. (1975)
Aquatic plants	1142	0.56	3.27	0.007	0.14	0.69	3.1	<0.1	5.24	0.16	0.09	36	Heffron et al. (1977)
Aspen, bark	140		2										A. J. Baker, Forest Products Laboratory, Madison, Wis., personal communication, 1978
Aspen, sulfite, parenchyma cell fines (unbleached)	670												Baker et al. (1975)
Aspen, sulfite screen rejects (mill 2)	66												Baker et al. (1975)
Aspen, whole tree		0.28					1.45	<0.05					A. J. Baker, Forest Products Laboratory, Madison, Wis., personal communication, 1978
Aspen wood	16												Baker et al. (1975)
Cage layer waste (processed)		1.25		0.3			0.2						Hamblin (1980)
		0.76					3.68						Helmer (1980)
		1.05					7.37						Helmer (1980)
		0.73					3.62						Helmer (1980)
		1.01					1.46						Helmer (1980)
		0.29					2.82						Helmer (1980)
							1.82						Helmer (1980)
Hardwood, mixed, screen rejects (mill 1)	520												Baker et al. (1975)
Hardwood, mixed, sulfite pulp fines (unbleached)	97												Baker et al. (1975)
Pine, southern, unbleached Kraft pulp fines (mill 5)	540												Baker et al. (1975)
St. Louis SCW-I	>1%	5.2	76				430	2.25					Belyea et al. (1979)
St. Louis SCW-II	>1%	5.1	220				520	2.80					Belyea et al. (1979)
Swine waste (processed)	54.4	1					12.1						Pearce (1975)

Table 3 (*Continued*)

	U	Yb	Sb	As	Br	Bo	V	References
Aquatic plants	0.29	0.23	0.23	0.60	28.4	75.1	2.5	Heffron et al. (1977)
Aspen, whole tree				<0.1				A. J. Baker, Forest Products Laboratory, Madison, Wis., personal communication, 1978
Broiler litter				6.2				Caswell et al. (1978)
Broiler litter (processed)				4.4				Blair and Herron (1982)
				34.3				Blair and Herron (1982)
Broiler waste (processed)				44				Trakulchang and Balloun (1975a)
				42				Smith and Calvert (1976)
Cage layer waste (processed)				4.6				Blair and Herron (1982)
				3.6				Blair and Herron (1982)
				1.5				Hamblin (1980)
				2.07				Helmer (1980)
				0.82				Helmer (1980)
				0.54				Helmer (1980)
				0.43				Helmer (1980)
				0.71				Helmer (1980)
				5.23				Helmer (1980)
					27			Shannon et al. (1973)
					18			Shannon et al. (1973)
					8			Shannon et al. (1973)
					24			Shannon et al. (1973)
					22			Shannon et al. (1973)
					19			Shannon et al. (1973)
					13			Shannon et al. (1973)

Table 3 (*Continued*)

	Ti	W	Ba	Cs	Rb	Sr	Ce	Dy	Eu	Lu	Sm	Si	Th	References
Algae												1.11		NRC (1971)
												1.73		Hintz et al. (1966)
Aquatic plants	91	0.9	41.9	0.2			2.0			0.14			0.46	Heffron et al. (1977)
Aspen, bark			85											A. J. Baker, Forest Products Laboratory, Madison, Wis., personal communication, 1978
Aspen, sulfite parenchyma cell fines (unbleached)			32											Baker et al. (1975)
Aspen, sulfite screen rejects (mill 2)			13											Baker et al. (1975)
Aspen, wood			19											Baker et al. (1975)
Hardwood, mixed, screen rejects (mill 1)			91											Baker et al. (1975)
Hardwood, mixed, sulfite pulp fines (unbleached)			21											Baker et al. (1975)
Pine, southern unbleached Kraft pulp fines (mill 5)			16											Baker et al. (1975)

TABLE 4 Essential Amino Acid Composition (dry basis)

Type of Resource	Crude Protein (%)	Arg (%)	His (%)	Iso (%)	Leu (%)	Lys (%)	Met (%)	Cys (%)	M+C (%)	Phe (%)	Tyr (%)	P+T (%)	Thr (%)	Try (%)	Val (%)	References
Algae	53.0	2.23	0.60	1.49	3.35	2.59	0.78			2.19	1.03		1.31	1.61	2.31	McDowell and Leveille (1963)
																Dam et al. (1965)
	61.3	3.98	1.09	2.20	4.41	3.50	0.81			2.65			2.02		3.01	Clement et al. (1967)
	59.3	2.81	0.94	3.70	4.92	2.81	0.84			3.05	2.42		2.80	0.86	3.98	Boyd (1973b)
	58.9	3.79	1.15	2.07	3.89	3.17	0.82			2.20	1.72		2.26		2.65	Chung et al. (1978)
	68.4	4.40	1.06	2.17	3.96	2.51	0.81			2.21	1.62		1.53	1.12	2.83	Anonymous (1970)
				4.20	6.01	3.12	1.76			3.06	3.22		3.55	1.0	4.55	
Aspergilis fumagatus	1–21	5.8	1.3	4.5	7.1	7.0	1.4	0.7		4.8			3.7		5.4	
Broiler litter (processed)	32.0	0.52	0.25	0.65	1.01	0.6	0.18	0.13		0.55	0.36		0.58		0.84	Bhattacharya and Fontenot (1966)
	30.6	0.58	0.24	0.64	1.0	0.58	0.13	0.09		0.54	0.33		0.57		0.82	Bhattacharya and Fontenot (1966)
	39.1	0.68	0.29	0.65	1.09	0.73			0.87			1.16	0.68	0.39	0.9	Blair and Herron (1982)
Broiler waste (processed)	45.2	0.6	0.22	0.5	0.87	0.55	0.13	0.18		0.51	0.38		0.62		0.81	Kubena et al. (1973)
	44.1	0.57	0.22	0.51	0.86	0.51	0.2	0.38		0.52	0.41		0.69		0.75	Kubena et al. (1973)
	44.9	0.67	0.26	0.58	0.94	0.66	0.22	0.39		0.57	0.46		0.71		0.79	Kubena et al. (1973)
	44.9	0.61	0.26	0.49	0.98	0.63	0.2	0.58		0.58	0.45		0.67		0.66	Kubena et al. (1973)
	42.0	0.68	0.24	0.5	0.95	0.61	0.19	0.39		0.55	0.47		0.76		0.67	Kubena et al. (1973)
	41.5	0.74	0.34	0.63	1.11	0.81	0.19	0.44		0.64	0.5		0.76		0.8	Kubena et al. (1973)
	45.3	0.78	0.35	0.59	1.05	0.81	0.21	0.64		0.67	0.52		0.8		0.77	Kubena et al. (1973)
	43.2	0.62	0.27	0.63	1.03	0.65	0.19	0.24		0.66	0.59		0.63		0.46	Kubena et al. (1973)
	47.8	1.53	0.3	0.68	1.31	0.82	0.5	0.51		0.77	0.77		0.83		0.82	Bhargava and O'Neil (1975)
	34.6	1.32	0.46	0.63	1.32	1.0	0.67	0.28		0.86	0.62		0.86		1.18	Bhargava and O'Neil (1975)

Cage layer waste (processed)		0.66	0.32	0.49	1.0	0.63	0.26	0.4		0.55	0.4		0.63		0.69	Parigi-Bini (1969)
		0.35	0.21	0.33	0.55	0.33	0.12	0.02		0.32	0.27				0.47	Hodgetts (1971)
		0.37	0.21	0.38	0.57	0.38	0.1	0.03		0.36	0.28				0.5	Hodgetts (1971)
	18.4								0.22							Shannon et al. (1973)
	31.6								0.29							Shannon et al. (1973)
	37.6								0.34							Shannon et al. (1973)
	31.9								0.35							Shannon et al. (1973)
	21.9								0.3							Shannon et al. (1973)
	38.5								0.51							Shannon et al. (1973)
	35.0								0.26							Shannon et al. (1973)
	38.3								0.67							Shannon et al. (1973)
	35.8	0.63	0.31	0.4	0.73	0.56	0.14	0.19		0.41	0.35		0.55		0.51	McNab et al. (1974)
	29.3	1.13	0.42	0.83	1.52	0.95	0.15	0.15		0.95	0.45		0.52		0.96	Stapleton and Biely (1975)
	31.3	0.45	0.21	0.41	0.67	0.44						0.7			0.58	Blair and Herron (1982)
	31.9	0.71	0.27	0.66	0.96	0.71			0.59			1.1	0.63		0.75	Blair and Herron (1982)
Cattle waste	20.3	0.18	0.12	0.21	0.62	0.47	0.09			0	0.03		0.29		0.38	Anthony (1971)
	18.4	0.16	0.1	0.23	0.49	0.37	0.07			0.07	0.23		0.23		0.32	Blair and Knight (1973)
Dried blood				0.9	11.9	8	1.3			6.8			4.4			Doty (1973)
Duckweed	41.1	4.29	1.89	3.87	7.15	4.13	0.83			4.45	2.91		3.20		4.96	Rusoff et al. (1980)
Duckweed																
Justicia americana	47.4	3.07	1.08	2.61	4.38	2.73	0.97			2.90			2.26		2.60	Boyd (1968)
	52.3	3.24	1.11	2.50	4.55	2.94	0.91			2.80			2.45		3.04	Boyd (1968)
	37.5	2.65	1.01	2.23	3.93	2.73	0.80			2.64			2.10		2.96	Boyd (1968)
Orontium aquaticum	49.6	3.17	1.02	2.32	4.30	2.64	0.84			2.80			2.26		2.55	Boyd (1968)
Nymphaea odorata	40.0	2.83	1.07	1.97	3.82	2.73	0.71			2.24			1.91		2.62	Boyd (1968)
Alternanthera philoxeroides	31.4	1.12	0.63	0.94	1.72	1.61	0.20			[a]			0.96		1.37	Boyd (1968)
Sagittaria latifolia	42.8	2.14	1.14	1.46	1.87	1.52	0.56			[a]			1.57		1.80	Boyd (1968)
Gliocladium deliquescens		5.17	2.33	4.06	6.18	6.15	1.19	1.42		3.96			4.86	2.31	4.85	Boyd (1968)
Meatpacking sludge	44.8	1.25	0.32	1.53	2.77	1.65	0.93	0.35		2.45			1.97		2.10	Paulson and Lively (1979)

TABLE 4 (Continued)

Type of Resource	Crude Protein (%)	Arg (%)	His (%)	Iso (%)	Leu (%)	Lys (%)	Met (%)	Cys (%)	M+C (%)	Phe (%)	Tyr (%)	P+T (%)	Thr (%)	Try (%)	Val (%)	References
Penicillium crysogenum		10.0	3.8	2.1	5.8	2.1	1.2	2.0		3.5			2.7	—	3.9	Rogers and Coleman (1978)
Potato *Aspergillus niger*	37	1.74	0.67	1.96	3.18	1.55	1.18	0.30		2.85			1.78	0.44	2.41	
Potato by-product meal		0.227	0.100	0.248	0.394	0.272	0.126			0.280			0.219			Macgregor et al. (1978)
Potato Symba yeast	51.1	2.86	1.02	2.20	3.83	3.22	0.77	0.51		2.76			2.76	0.66	2.15	Skogman (1976)
Streptomyces fradiae		1.3	—	2.68	—	1.98	1.32	—		—			3.23	—	3.5	
Streptomyces rimosus		4.79	4.2	1.76	4.87	1.0	—	—		2.37			—	—	5.54	
Streptomyces viridifaciens		1.93	—	1.71	—	—	0.62	1.09		2.89			2.19	—	9.48	
Swine waste (processed)	24.5	0.33	0.1	0.29	0.51	0.32	0.13			0.23	0.29		0.39		0.43	Harmon et al. (1972)
	31.8	0.74	0.28	0.51	0.92	0.72	0.22			0.72	0.56		0.67		0.75	Harmon et al. (1972)
		1.28	0.47	1.49	2.79	1.42	0.77				1.17		1.96	0.28	2.06	Harmon et al. (1973)
		0.67		1.03	1.57	1.11	0.77			0.87			0.8			Orr et al. (1973)
	23.5	0.58	0.29	0.73	1.48	1.02	0.38	0.3		0.8	0.65		0.78		0.85	Kornegay et al. (1977)
		0.57	0.27	0.65	1.17	0.93	0.35	0.35		0.72	0.57		0.82		0.75	Kornegay et al. (1977)
Tomato seeds	24.5	2.16	0.54	0.86	1.44	1.21	0.19	0.15		0.89			0.74	0.23	0.91	Tsatsoronis and Boskou (1975)
Tomato skins	10.0	0.398	0.146	0.278	0.506	0.44	0.075	0.049		0.308			0.467		0.5	Tsatsoronis and Boskou (1975)
Tomato seed cake	39	3.38	0.77	1.53	2.22	2.11	0.65	0.31		1.64			1.25	0.394	1.54	Ben-Gera and Kramer (1969)
Trichoderma virde		2.98	1.67	3.52	5.35	3.94	1.20	1.38		2.76			3.94	1.80	4.48	
Verticillium sp.		—	—	4.0	5.8	4.7	1.1	—		3.5			3.8	—	4.9	

*a*Trace.

TABLE 5 Nonessential Amino Acid Composition (dry basis)

Type of Resource	Aspar A (%)	Ser (%)	Glut A (%)	Prol (%)	Gly (%)	Alan (%)	Cys (%)	References
Algae	2.82	1.04	3.69	2.19	2.01	2.17	0.18	McDowell and Leveille (1963)
	5.27	2.57	7.72	2.39	2.91	4.17	0.25	Clement et al. (1967)
	5.54	2.26	6.07	1.98	2.75	3.94	0.42	Boyd (1973b)
	3.19	1.44	4.90	1.80	2.25	3.00	—	Chung et al. (1978)
	5.22	3.68	8.82	2.43	3.35	5.29	0.65	Anonymous (1970)
Duckweed	7.12	2.61	7.60	2.93	3.79	4.59	—	Rusoff et al. (1980)

TABLE 6 Vitamin Composition of Algae (mg/100 g)

Type of Resource	Carotene	Ascorbic Acid	Thiamine	Riboflavin	Niacin	Vitamin B_6	Pantothenic Acid	Folic Acid	Biotin	Vitamin B_{12}	References
Algae	60.2	39.6	1.15	2.69	10.8	1.10	4.60	72.9	0.168	0.107	Cook (1962)
	—	13.0	0.11	2.31	10.8	0.13	1.66	—	—	—	McDowell and Leveille (1963)
			40.5	37.5	107.0	3.1	8.9	0.45		1.55	